INTELLIGENT MULTIMEDIA COMPUTING SCIENCE

Business Interfaces, Wireless Computing, Databases, and Data Mining

INTELLIGENT MULTIMEDIA COMPUTING SCIENCE

Business Interfaces, Wireless Computing, Databases, and Data Mining

Cyrus F. Nourani
University of California,
Los Angeles, California, USA

AMERICAN SCIENTIFIC PUBLISHERS
25650 North Lewis Way
Stevenson Ranch, California 91381-1439, USA

AMERICAN SCIENTIFIC PUBLISHERS
25650 North Lewis Way, Stevenson Ranch, California 91381-1439, USA
Tel.: (661) 254-0807
Fax: (661) 254-1207
E-mail: order@aspbs.com
WEB: www.aspbs.com

Intelligent Multimedia Computing Science: Business Interfaces,
Wireless Computing, Databases and Data Mining by Cyrus F. Nourani

The cover is designed by Dr. Cyrus F. Nourani

This book is printed on acid-free paper. ♾

Library of Congress Control Number: 2004114448
International Standard Book Number: 1-58883-037-3

PRINTED IN THE UNITED STATES OF AMERICA
10 9 8 7 6 5 4 3 2 1

Preface

Intelligent Multimedia Computing Science is an interdisciplinary field combining the arts, sciences, artificial intelligence, computer science, mathematics, and the humanities. The areas present new challenges to artificial intelligence (AI), computing. There are applications presented to business intelligence, intelligent databases, intelligent datamining, knowledge representation (KR), spatial computing, and wireless computing interfaces. The scientific basis to intelligent multimedia computing inter-twined with the knowledge representation areas that embody the science are the basis. Intelligent Multimedia Sciences is the study, invention, and creative use of sciences and technologies for understanding and expression of multimedia intelligence and its computation by people and machines. The field presented is deeply rooted in AI, mathematical logic and models, modern communications, computer, and humanities. Academic digital media studies are at times a partnership among Arts and Sciences, Computer Science, and Mathematics. The academic programs must intimately become aware as to the interdisciplinary nature of intelligent multimedia and its range of basic research within an applications-oriented environment. The new fields encompass the Intelligent and cognitive aspects of media Arts and Sciences, exploring the technical, cognitive, and aesthetic bases to human multimedia intelligence and its computation, the applications to business intelligence, model discovery, data mines and intelligent data bases, and information technology (IT). The monograph is a technical and practical book to the popular audience, to the business minded professionals, and to all groups wanting to be on an intelligent basis to the new field. Academics and students with sense of enterprise and inquiry will be lead to develop their own new paths. Since research is an important component of academic programs, the book provides new horizons toward specific areas of research interest. The monograph can be a university book to a course on intelligent multimedia sciences and all professional career paths within it. It is further a book for the professionals, the business intelligence, data mining, IT career persons, and the popular science enthusiasts. Each chapter has assigned exercises for the academic applications.

The technical overview is as follows. There are several Knowledge Representation, abbreviated KR, techniques applied on the various chapters to carry out intelligent multimedia. The techniques range from basic model diagrams and predicates, nets, and frames, to agent visual objects, agent game trees, graphical casts, spatial groupings, splines, macrodot abstractions, virtual trees, and agent hypermedia active windows. The computing techniques, a language, the Morph Gentzen deductive system and its models are presented in brief. Basic application areas we start with as examples are designing predefined visual scenes with diagram composition and combination. The application area is based on AI planning include affective computing. Reasoning and planning can be applied to define dynamics based on descriptions and compatibility relations. Business Intelligence, predictive modeling, and multimedia data bases and data mines are important new technical areas presented on several chapters. The mathematical basis presented on two chapters and spread on the book with agent computing techniques.

There are techniques presented with television program rating prediction possibilities. The new Intelligent Language paradigm and intelligent visual computing paradigms are applied to the intelligent multimedia (IM) multiagent multimedia computing. Multimedia objects are viewed as syntactic objects to which a deductive system is applied. There are morphisms defined where a Gentzen deductivc, Morph Gentzen, system is put forth for Intelligent Multimedia computing with the morphisms. The application areas to Animation, Motion Pictures design, and virtual Reality are presented with new innovative techniques.

There are further computing application areas to Visual representations for motion interpretation, inferring context, high-level models of action and intention, plan recognition given perceptual sensing, learning and recognizing procedures from video. Business applications to stock forecasting and portfolios are outlined. Active databases deploy certain computing that lends itself naturally to the IM principles.

Intelligent objects embed the concepts of an active object and "events" are embedded by the computing defined by Intelligent Multimedia as the starting principles. Intelligent Multimedia Data Mining Applications are presented. Specific design techniques and paradigms are presented for most areas with the author's innovations.

Applying the Morph Gentzen computing logic is a way to model discovery and a basis for a mind computing model in our projects. The influence of intelligent multimedia is bound to increase, and to profoundly change how we communicate with the world around us and how we express ourselves via computers and in person. It will further influence how we design movies and television programs. The principal technical and design challenges of the emerging information age are the development of the academic, scientific, and practical basis to carry out intelligent multimedia.

The computing and internet technologies have increased the rate at which information is conveyed. Knowledge representation has two significant roles: to define a model for the AI world, and to provide a basis for reasoning techniques to get at implicit knowledge. Technology has begun presenting artificial worlds at virtual reality, where we can 'navigate' as in illusions. Augmented reality is a new area to which chapters in the book contribute. Spatial computing and imaging is an important area to the author since 2000 and the several new multimedia computing groups. The mathematical intelligent object basis has been applied to the author presents can be applied to. Intelligent Multimedia Datamining is an important application area presented with new specific techniques. Other areas are of interest in knowledge management include metadata or manual annotation of the content, which may prove useful in identifying the document content or adding additional information beyond the content itself. The abstract visual computational linguistics and context abstraction techniques presented in the book can be applied to realize the goals above.

Inevitably, the visual interface will become three-dimensional as more realistic and complex tasks are taken on. New 3-D media is being invented and studied to make complex structures more readily understandable. The Interactive Film is on areas where digital distributed networks are transforming the structure and language of the cinematic experience. Future stories will be multithreaded and adaptive, responding and learning from interactions with their audience as well as with their authors. Gesture and narrative languages group studies how artifacts (such as agents and toys) can be designed with psychosocial competencies based on a deep understanding of human linguistic, cognitive, and social abilities. The responsive multimedia works often involve a collaboration among many people of varying expertise and require innovation at many levels, from the incorporation of a new sensor or display technology to the structuring of media objects and their behaviors. The boundary between the content of information and its physical representation is another study area. Personal Information Architecture has to do with: What does it mean to put digital computing and communication into furniture, clothing, housing, toys, and just about everything around us. Intelligent multimedia might be a way to a serene scientific mind, while not closed to the phantasm offered at virtual worlds, allowing us to discover new thoughts. Intelligent Multimedia techniques and paradigms are defined. Basic multimedia computing and principles are presented for multimedia paradigms with many facets and application areas.

A computational understanding of the link between visual and textual information is necessary so that media processing technologies can be integrated to extract and combine rich representations of media content at a high level. Such representations facilitate more intuitive and personalized access and navigation of media artifacts for users, and facilitate knowledge management services such as knowledge mining. On content retrieval applications textual information from TV sit-com scripts was combined with visual features, through user interaction, in order for a system to locate scenes containing a particular character. Abstract planning, models, and virtual trees planning are presented and applied to the disciplines presented. Another area presented is Intelligent Cyberspace Interfaces and Wireless Access Protocol generic mobile Computing. Many technologies can form components of an intelligent knowledge management system for unstructured data. These may include the following: information retrieval, attempting to satisfy a user's information need by locating documents containing relevant information; summarization, content reduction by specialization or generalization of a document; information extraction, locating facts from within a document and potentially inferring other information; machine translation, translation of content between one or more languages; content indexing, speech or image recognition and analysis to identify the content of multimedia data; and information visualization, to represent the contents of information sources or their interrelationships.

The author's own first start to practical multimedia applications to what he had called Intelligent multimedia and the business applications, double vision computing project 1995, had to do with practical personal information and marketing E-commerce. There are business schools engaged on the areas since the author's 1998 projects. UCLA's Center for Digital Media established within the Anderson Graduate School of Management, addresses the business and management issues of digital media. The accelerating pace of development and innovation in these areas will transform existing industries, alter the workplace, and bring new business into being and, in the process impact every aspect of business. The Center's studies address these timely issues. The book presents new techniques to address business intelligence interfaces with multi-tier designs.

My thoughts and personal gratitude to Ramna, CamNieman, Fraunhofer InstitutMedienkommunikation, Sank Augustin, Germany. To Ina, ProDesign and Marketing Berlin, Christian and Claudia. A+Morph, Berlin. To the colleagues from at least three disciplines for being patient and encouraging with accepting all my papers to conferences and journals at difficult areas and times. To my family for being patient with my perhaps hard to understand dacademic life and state of mind. To Dr. H. S. Nalwa and American Scientific Publishers.

My profound gratitude to Tech-Net Scientific e-Publishing Services and David Tebera for their superb publishing and attitude.

Cyrus F. Nourani
University of California,
Los Angeles, California, USA

Contents

Chapter 4: Basic Intelligent Multimedia Techniques

Chapter 5: Agent Computing and Intelligent Multimedia

Chapter 6: The Mathematical Foundations

Chapter 7: Intelligent Visual Databases and Data Mines

Chapter 8: The Applications to Business and E-Commerce

Chapter 9: Spatial Heterogeneous Visual Computing

Chapter 10: Knowledge Representation and Predictive Intelligent Multimedia Diagram Models

Chapter 11: Business Intelligence, Content Exploration, and Intelligent Interfaces

Chapter 14: Intelligent Cyberspace Interfaces and Wireless Access Protocol Generic Computing

About the Author

Dr. Nourani received his BS, MS Science, and Ph.D. degrees from MIT and UCLA. Academic appointments include CS, management science & IT, and mathematical sciences. He has an international reputation in the areas of computer science, AI, information systems, decision sciences, EC, ERP, IT, predictive modeling, software, and programming methodology and many years of experience in design and implementation of computing systems, since university freshman year. Dr. Nourani had ventured into intelligent multimedia since 1994 in California and created Morph Gentzen computing logic. There are multimedia projects announced with Fraunhofer Institue Medienkommunikation, Sank Augustin, Germany. He started his R&D career at IBM Research in Yorktown Heights and has many years of R&D experience at major computing and aerospace companies. Dr. Nourani's academic experience includes faculty positions at the University of Michigan, Ann Arbor, the University of Pennsylvania, Philadelphia, the University of Southern California, UCLA, and MIT and visiting professor positions at Australian Universities ECU Perth, the University of Auckland Management Science, and the University of California, Santa Barbara. He began independent consulting with Aerospace companies in 1986. As part of the research staff, he has designed and developed AI robot planning, autonomous flight control systems, and AI reasoning systems at Northrop Research and Technology Center, California, and at GE Astropace Systems. He also has several years of AI, software, and computing foundations R&D experience at GTE Research Labs. In addition, he has taught AI to the Los Angeles aerospace industry and has coauthored many R&D and commercial ventures. Over the years, he has held academic membership in the New York Academy of Sciences, AAAI, ACM, AMS, ASL, IEEE, and the EATCS-European Association Theoretical CS. Over 150 publications in areas ranging from computer science, management science, and mathematics to cognitive science, computational linguistics, logic, and perception are credited to him.

CHAPTER 1

Intelligent Multimedia

CONTENTS

1. INTRODUCTION

1.1. The Area

Intelligent Multimedia is an interdisciplinary field combining the arts, sciences, artificial intelligence (AI), computer science, mathematics, and the humanities. The areas present new challenges to AI, computing, and knowledge representation (KR). The scientific basis to intelligent multimedia computing inter-twined with the knowledge representation areas that embody the science are presented. The computing and internet technologies have increased the rate at which information is conveyed. Knowledge representation has two significant roles: to define a model for the AI world, and to provide a basis for reasoning techniques to get at implicit knowledge. There are several KR techniques applied on the various chapters to carry out intelligent multimedia. The techniques range from basic model diagrams and predicates, nets, and frames, to agent visual objects, agent game trees, graphical casts, spatial groupings, splines, macrodot abstractions, virtual trees, and agent hypermedia active windows. Intelligent multimedia science might provide glimpses to a serene scientific mind, while not closed to the phantasm offered at virtual worlds, allowing us to discover new thoughts. Technology has begun presenting artificial worlds at virtual reality, where we can 'naviagate' as in illusions. A new area the book contribute to is Augmented reality.

There are research centers with staff devoted to virtual and augmented reality. Co-operation with theatre, dance, film, television are amongst the areas supported by the new disciplines, with new productions. Filmmakers, producers and distributors are introducing digital technology to motion picture production, business planning and marketing. Content management is a central area treated in the present book. Is the 1960s cybernetics community prediction that machines would be smarter than people within 20 years a reality? Artificially intelligent computers are applied to building design, travelling to space and solving critical crises decisions. There are also a nice pets, robot butlers, and nannies to watch the kids.

The growing use of digital technologies in film/TV is opening the door to a variety of exciting new business opportunities for media and entertainment companies. During the last three years, in particular, IBM has focused its efforts to providing digital technology

ISBN: 1-58883-037-3/$35.00

Intelligent Multimedia Computing Science
By Cyrus F. Nourani
Pages: 1–11

solutions to the entertainment industry. Experienced entertainment industry executives are hired to enhance its renowned technology expertise. Working with the global entertainment and broadcast companies, industry-specific solutions are being developed.

The areas explored are: Dialogue Systems; Language Modeling; Information Processing; Multimedia Applications Interfaces; and Visual Communications. The academic areas are being explored around the world. A particular important university media laboratory is MIT with major academic disciplines since the last few years. The author has pursued what he called Intelligent Multimedia since 1995 as a new scientific discipline, introduced here, apparently not disjointed from the areas engaged on at the above specific organizations. The monograph might provide the multimedia enthusiasts and students with a practical and academic basis to mingle, in the intellectual and professional senses, with such interest groups.

Amongst the new area explored are Affective computing based on an emotion models to intelligence pursued at R. Picard's group at MIT Media, and contributed to a degree here at Intelligent Multimedia. The News in The Future group contends providing tools for the creation of and access to content is not sufficient to guarantee effective communication. Effective communication requires distribution of messages that are rich in description. These descriptions, when taken with confederation of the context of both individuals and communities receiving them, can result in messages that are both engaging and useful.

Spatial computing and imaging is an important area to the author Nourani (2000) and the several new multimedia computing groups (Bailey et al. 2004). The mathematical intelligent object basis has been applied to the author presents can be applied to. Spatial objects are applied with equivalence classes and classes to create abstract knowledge descriptions in the author's projects since 1994. Bailey et al. (2004) are further illustrations and progress on the areas, where specific representational primitives (a spatial "vocabulary") and inferencing for these tasks are presented.

Inevitably, the visual interface will become three-dimensional as more realistic and complex tasks are taken on. New 3-D media is being invented and studied to make complex structures more readily understandable. The Interactive Film is on areas where digital distributed networks are transforming the structure and language of the cinematic experience. Future stories will be multithreaded and adaptive, responding and learning from interactions with their audience as well as with their authors. Object-Based Media explores the creative and technological applications and implications of audio and video communication systems in which intelligent processes "understand" the inputs, producing a scene description in terms of constituent objects and scripting information.

Gesture and narrative languages group studies how artifacts (such as agents and toys) can be designed with psychosocial competencies based on a deep understanding of human linguistic, cognitive, and social abilities. Current Sociable Media explores new directions in the design of virtual social environments. The questions are relating to identity and society in the networked world: How do we perceive other people on-line? What might a virtual crowd look like? How can the underlying technology shape the evolution of on-line culture? Another area is called Responsive multimedia, referring to a melding of media forms that has the ability to react in some way to audience identity, circumstances, environment, and actions.

The responsive multimedia works often involve a collaboration among many people of varying expertise and require innovation at many levels, from the incorporation of a new sensor or display technology to the structuring of media objects and their behaviors. Tools are being designed to support the multi-level nature of such projects, in terms of both authoring and user-base. Another area is building interactive animated characters for use in virtual environments such as immersive story-telling systems, games, and web-based worlds. Research issues include: the role of emotion and motivational state in action-selection, learning and adaptation, directability (i.e. the integration of external control with autonomous behavior), synthetic perception, and models of motor control for expressive movement.

The Tangible Media explores the Tangible Bits vision to seamlessly couple the dual worlds of bits and atoms. Human interfaces are being designed that employ physical objects, surfaces, and spaces as tangible embodiments of digital information and processes. The boundary between the content of information and its physical representation is another study area. Personal Information Architecture has to do with: What does it mean to put digital computing and communication into furniture, clothing, housing, toys, and just about everything around us.

The author's own first start to practical multimedia applications to what he had called Intelligent multimedia and the business applications, double vision computing project (Nourani 1995), had to do with practical personal information and marketing e-commerce (EC) around 1996. The practical areas on language visualization is side benefit application area to multimedia computing. Practical multimedia literature is not as abundant as might be perceived. There are business schools engaged on the areas since the author's 1998 projects. University of California Los Angeles's (UCLA) Center for Digital Media established within the Anderson Graduate School of Management, addresses the business and management issues of digital media. The accelerating pace of development and innovation in these areas will transform existing industries, alter the workplace, and bring new business into being, and, in the process impact every aspect of business. The Center's studies address these timely issues.

Are the 1960s cyberneticians predictions that machines would be smarter than people within 20 years a reality? Artificially intelligent computers are applied to design building, travel to Mars and solve decision crises on earth. There are also pets, and robot butlers created or robot nannies to watch the kids. But somewhere between promise and production, the fantasy got confused. AI came to mean to the public a computer that could beat most chess players. Intelligent multimedia is clearly not upon us to computerize chess games. A multiagent-chess games was however, a specific artificial intelligence paradigm that the author had put forth.

There are projects investigating the synthesis (Salway et al. 2003) of language technologies, like information extraction and corpus-based text analysis, video data modeling and knowledge representation. The aim is to develop a computational account of how video and text can be integrated by representations of narrative in multimedia systems. The multimedia domain is that of film and audio description—an emerging text type that is produced specifically to be informative about the events and objects depicted in film. Narrative might be an important concept for intelligent multimedia knowledge management. Story understanding systems have focused typically on relatively short and simple written texts, like newspaper stories and children's stories. However, more elaborate scenarios are how stories are conveyed and understood in a moving image and in spoken or written language. The relationship between different kinds of media is also required for the generation of enriched multimedia information sources.

A computational understanding of the link between visual and textual information is necessary so that media processing technologies can be integrated to extract and combine rich representations of media content at a high level. Such representations facilitate more intuitive and personalized access and navigation of media artifacts for users, and facilitate knowledge management services such as knowledge mining. The need to integrate media representations, at different levels of abstraction, leads to the idea of a media interlingua (Maybury 1997). For example, low-level integration may involve representations of pixel distributions, sound wave features and keyword frequencies that can be expressed numerically. Higher level 'semantic media content' is equated with named events, objects, their attributes and their relationships. Various logics have been proposed in the multimedia literature and the present book contributes to an emerging science applying annotations and context abstraction to a visual computational linguistics. However, there are higher levels still, which are not yet dealt with in multimedia systems. For example, it may be necessary to represent sequences of connected events, event causal relationships and the intentions for potential applications. On content retrieval applications textual information from TV sit-com scripts was combined with visual features, through user interaction, in order for a system to locate scenes containing a particular character (Wachman and Picard 2001).

Many technologies can form components of an intelligent knowledge management system for unstructured data. These may include the following: information retrieval, attempting to satisfy a user's information need by locating documents containing relevant information;

summarization, content reduction by specialization or generalization of a document; information extraction, locating facts from within a document and potentially inferring other information; machine translation, translation of content between one or more languages; content indexing, speech or image recognition and analysis to identify the content of multimedia data; and information visualization, to represent the contents of information sources or their interrelationships.

Other areas are of interest in knowledge management include metadata or manual annotation of the content, which may prove useful in identifying the document content or adding additional information beyond the content itself. The abstract visual computational linguistics and context abstraction techniques presented in the book can be applied to realize the goals above.

1.2. Intelligent Multimedia (IM) Knowledge Representation

Knowledge representation is a cornerstone to AI and has roots in first order logic (FOL) inheritance hierarchies, frames, and semantic nets. With the emergence of multimedia computing there are new facets and dimensions to KR. The starts to KR were on several basic and more abstract directions. Semantic-nets are knowledge representation techniques which specify inheritance, a concept which is since automatic in object computing and object paradigm KR (Brachman et al.). Semantic nets can be specified with abstract data types and automatic implementations realized (Nourani-Lieberherr 1984) where abstract objects are specified with axioms and operations. Inheritance hierarchies are taxonomies organised by the usual Is-A relation, as in Figure 1.

Intelligent behavior or computing is based on the knowledge a computing agent has about its environment. Most knowledge can be presented in declarative form describing the world conceptualisation with objects and their interrelationships, for example. The interrelationships are managed with data and knowledge bases.

Object := Coffee_Constellation
OPS := Serve_Coffee (Type, Table_no) | . . . . . .
Serve_Coffee (Spectacular_Brew, n)
    ⇒ Signal an available robot to fetch and serve (Spectacular_Brew, table n)
Exp := Serve_Coffee (Angelika, Table_no) | . . .
Serve_coffee(Angelika,Table_no) ⇒ if out_of_Angelika notify Table_no;
        offer cookie⟨and make use of
        intelligent decision procedures to
        offer alternatives⟩

Patterns, schemas, AI Frames and viewpoints are the micro to aggregate glimpses onto the database and knowledge bases were masses of data and their relationships-representations, respectively, are stored. Schemas and frames are what might be defined with objects, the object classes, the object class inheritances, user-defined inheritance relations, and specific restrictions on the object, class, or frame slot types and behaviors.

A scheme might be
Intelligent Forecasting

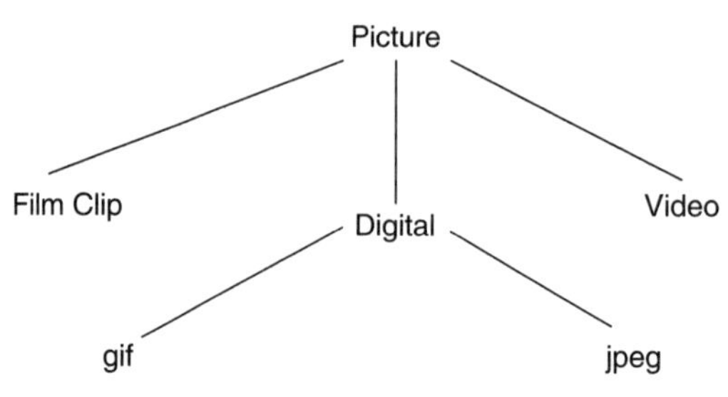

Figure 1. Inheritance hierarchy.

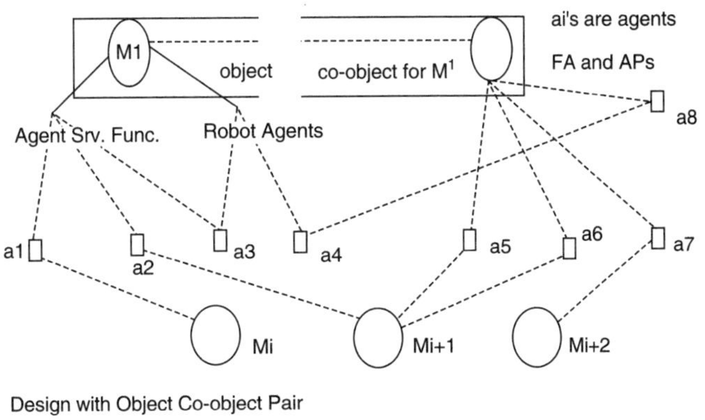

Figure 2. Multiagent object knowledge representation.

IS-A Stock Forecasting Technique
Portfolios Stock, bonds, corporate assets
Member Management Science Techniques

Schemas allow brief descriptions on object surface properties with which high level inference and reasoning with incomplete knowledge can be carried out applying facts and the defined relationships amongst objects. Relationships: Visual Objects A and B have mutual agent visual message correspondence. Looking for patterns is a way some practical AI is a carried on with to recognise important features, situations, and applicable rules. Further visual objects and specifics are how knowledge is represented as depicted on the following example from the chapters on the book. The book presents the new KR areas starting with multiagent visual objects, hybrid pictures, agent tress, agent game trees, to nurb splines, agent frame, and movie and Morph templates. There are XML, panels on: Instance, Frame, Effect, Sound, Stroke, Fill, Text, for example. Starting from FOL where we were on AI, looking back, we have to lift and create the new science to multiagent multimedia KR to carry onto the intelligent multimedia computing arena. The task has in part been carried on since 1990's on the author's projects and a new multimedia computing multitude. FOL KR is based on a formal system, a structure built to embody a logician's theories. Many-sorted structures are represented and the notion of a partial model is introduced to represent knowledge in explicit terms, while a reflection principle associates a theory with a meta-theory to carry out deductions.

At FOL or inheritance hierarchy KR at issues is what world is being represented. For example, the world might be depicted from Figure 3. Thus we can derive with ordinary deduction extensions to the knowledge derivable. A preliminary categorical and logical counter-parts to the above has a start with the Clear specification paradigm (Burstall and Goguen 1977) on the specification fronts, and the authors model-theoretic (Nourani 1980) algebraic computing on the deductive areas since 1980's. The latter projects represent structural knowledge with algebraic logic, models, and categories. With intelligent multimedia

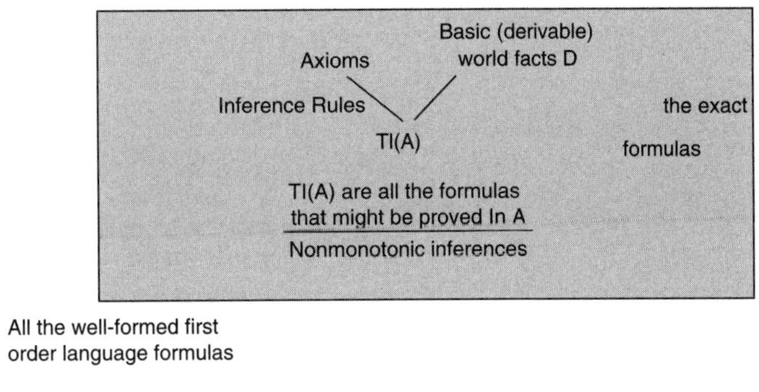

Figure 3. The FOL world.

KR the above world logical view in Figure 1 is on a multiagent pictorial level with the items on Figure 1 occurring on a meta-descriptions a carried on with agent computations. The structure A above might only be a virtual and flexible amorphous universe where a logic is carried on virtual trees with virtual leaves. The logic that is ascribed to a frame computation occurs on higher levels with hybrid pictures with multiagents with virtual trees and a Morph Gentzen deductive system on structures that are not as rigid as a frame with flexibility upto agents and pictorial specifics and aggregations.

A pictorial representation of a Frame is

Frame Name			
Parent	Frame parent 1		
Parent	Frame parent 1		
Slot	Slot1-name	Slot1-value1	
If-added	predicate-1a	If needed	predicate 1-b
Slot:	Slot2-name	Slot-2 value2	
If-added	predicate-2a	if-needed	Predicate 2-b

However, hybrid pictures and multiagent morph computing applies a higher level KR, depicted, for example from Chapter 2, agent trees from Chapter 3 on, or Splines as follows.

Member ⟨file edit, view, Insert, Modify, Control, Xtras, Window, help⟩

Toolbar

Score Window	A Frame-frame record of your Director Movie
Cells	The score is composed of cells, the storage unit that holds all the information about your movie. The columns of the cells are organized into frames. The rows make up the score channels.
Cast Window	The storage place for cast members in the Director. A cast window is organized into rows and columns of small windows, each containing a unique cast member.
A Cast Member	A multimedia component, e.g. bitmaps, sound
Control Palette	Processes VCR type control over play backs of your movie, rewind, play back. etc.
The Stage	Background at which your movie plays.
Paint Window	Built-in paint programs

Some standard 3D imaging techniques are applied by systems designed since 1998, for example 3D Studio MAX, is morphing where the shape of one object is changed over time. What is applied to get to VR: 3D Computer Graphics; modeling objects, data gloves, a see through HMD designed at MIT, 1982. Since then, there has been computer designed VR hardware and software. Dynamic objects are applied with collision detection avoidance at VR. A sprite is created as soon as you drag a graphical cast member from a cast window onto the stage, on into the score. Its function is that you can change its properties, size, blend percentage and which ink effect it uses; and the marginal cast member it is based on will not change. It allows you to vary based on the source cast member. To create Morphs agent trees and objects area applied to represent knowledge.

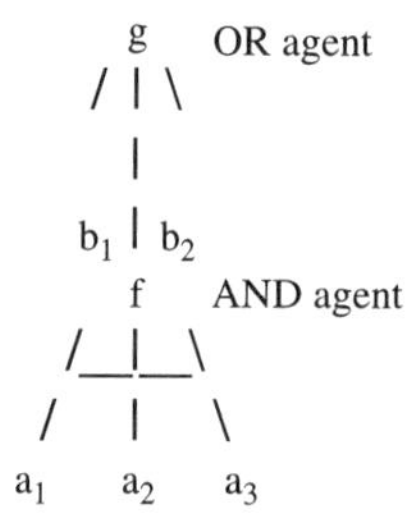

Figure 4. Agent and/or Trees.

For example chess game trees can be defined by agent augmenting AND/OR trees (Nourani 1997) and applied to economic games (Nourani 2000, Ch. 8 and 11) since and/or trees from Nilsson (1969).

Morphing applies KR with objects, hybrid pictures, which are certain agent-augmented pictures, presented on the book chapters. For example, the Morphs apply objects as follows:

Object Name:
Number of Vertices:
Min X =
Min Y = 2.0200014E−4, Max Y = 0.75251697
Min Z = −0.065170003, Max Z = 0.072493999
Number of Texture Vertices:
Number of Normals:
Number of Faces: 17535

Falsh MX improves the ActionScript robots programming language and enhanced visual designer with animator. Flash has further an animation program for the Web and simple application development. A Flash designer might apply XML creating custom objects in addition to using motion tweens. Splines are further example KR techniques applied.

Reasoning about spatial data is an essential task in many applications. The manipulation of qualitative spatial representations, for example, to detect whether one object will soon occlude another in a digital image or efficiently determine relationships between a proposed road and wetland regions in a geographic data set.

The diagrammatic KR the author has devised since 1980's is an abstract KR techniques where specific functions describing a world are applied as the essential basis to characterize with axioms and defining relations what is true in a world and to compute the logical extensions with the functional consequences as in Figure 2. Diagrammatic KR permeates the chapters as essential techniques applied with or as a higher abstraction to the above KR techniques or as an encoding. The diagram functions are applied to model discovery, model computations, and context abstraction on several areas presented in the book. Furthermore, the author had proved that diagrams encode the 'Possible Worlds' knowledge with minimal functions. That is an important achievement on the computational frontiers over a decade ago.

The secondary application levels are the areas for intelligent multimedia knowledge management technologies. One of the most compelling is the development of systems to enhance the effectiveness and availability of learning and study environments with audio-visual recording equipment and the ease with which recordings can be made available on-line via the World Wide Web. The technologies are increasingly being utilized for distance learning by students.

One important feature of recordings of this type is that they often contain tacit knowledge. Information delivered by speakers will often be unplanned, particularly when answering questions or taking part in discussions. The technology used to make these recordings varies. Another secondary and important is indexing audio and video content. The hard question for which there is not a simple answer in the technologies outside the present book is what should be indexed? Ideally, all aspects of the image should be captured, but what are the essential aspects? Examples are to identify the type of location and perhaps uniquely identify it. We could attempt to recognize all the objects in the scene and go further to track their motion in the scene. How should an object to classified? Should a car be classified as a car or a vehicle, and need it include attributes such as color, manufacturer, model, etc? At present there seems to be little work in the development of standardized image label taxonomies. Of course, taxonomies of this type will be of little value if robust feature extraction is not available. These questions all lead us to the KR areas.

Automatic Knowledge Management is one interesting feature in image analysis beyond classifying the natural scene is to recognize overlaid text. This material is often added to news broadcasts and other documentary material to identify locations or speakers. Optical character recognition (OCR) techniques can be adapted to video images to identify this text which can then be used to help label names of speakers or locations. Education is a potentially important application area for intelligent multimedia knowledge management. One source of knowledge is archived recordings of audio presentations including lectures and seminars. Recordings of presentations often contain multiple information streams involving visual and audio data. These areas are addressed in Jones and Edens (2003).

1.3. Organization of the Chapters

The monograph chapters contain three most technical Chapters 5, 6, and 10, which the non-technical persons can take at their strides. Chapters 6, 10, and 12 are mathematical and only necessary as foundations for the pure scientific applications. Chapter 2 is an overview to practical multimedia programming based on the available techniques and tools. Creative techniques are examined and the applications to animation, movies, video and virtual reality are presented. What is here does not apply AI yet. It is only the basic programming areas. The following chapters carry on Intelligent Multimedia with AI. Chapter 3 starts with basic AI introducing AI agent computing.

Multimedia AI systems are presented necessitating aesthetics with world models. Multimedia context computing and context abstractions are introduced leading into visual object perception dynamics. Intelligent multimedia hybrid pictures and hybrid morphing is presented. Next, a high-level multimedia language is defined and its applications outlined. OOP and IOOP—Intelligent OOP are introduced with visual objects leading onto Virtual Visual JAVA. Agent computing with mobile visual objects concludes the chapter. Chapter four starts with visual context and objects and multiagent intelligent multimedia. Context abstraction and meta-contextual reasoning is introduced as a new field. Multiagent visual multi-board planning is introduced as a basis to intelligent multimedia with applications to spatial computing. Visual model-based computing is introduced with the new techniques.

Chapter 5 introduces agent computing with interactive intelligent multimedia. An overview to a practical agent computing model based on beliefs, intentions, and desire is presented and possible augmentation to intelligent multimedia is explored. Dynamics and situation compatibility in introduced as a structural way to compute and compare epistemic states. Worlds, epistemics, and cognition for androids are introduced with precise statements.

The foundations are applied to present a brief on Computational Illusion, affective computing, and Virtual Reality. The IM_BID design model is presented. KR for AI Worlds, and Computable Worlds are presented with diagrams. Cognitive modeling is briefed with an introduction to ordinal dynamics. A preview to computational epistemology with cardinality and concept descriptions is introduced. Deduction models and perceptual computing is presented with a new perspective. Intelligent multimedia interfaces are an important component to the practical computational aspects.

Chapter 5 introduces practical Morph Gentzen and MIM—The IM Morphed Computing Logic for computing for multimedia are new projects with important computing applications. KR techniques with G-diagrams and applications to define computable models and relevant world reasoning. G-diagrams are diagrams defined from a minimal set of function symbols that can inductively define a model. Chapter 6 is on the mathematical basis to intelligent multimedia developed for the most part by the author since 1994. Chapter 7 on the scientific and practical basis to intelligent visual databases, TAIM to be specific, and the applications to data mining. The existing academic pursuits in the emerging area are explored. Intelligent active databases are introduced. The video technologies are examined on the specific video compression, storage, and retrieval areas. Digital technologies for providing industry-specific solutions and services to telecommunications, cable TV, wireless, broadcasting, and entertainment industries, as well as internet service providers are presented in brief.

Chapter 8 is on the business and commerce applications. Management Science, stock forecasting, Intelligent business objects and Intelligent Multimedia applications to IT are defined. Amongst the areas are agent computing, modern portfolios, heterogeneous computing, business objects, intelligent databases, intelligent objects, and decision science. A new view to management information systems (MIS) with agent computing and intelligent multimedia is presented. The new agent computing theories we have defined since 1994 can be applied to present precise strategies and prove theorems on games. The game is viewed as a multiplayer game with only perfect information between agent pairs. Enterprise modelling is presented based on the management process model. Innovation-market response criteria imply new areas for management science and economics games paradigms starting with heterogeneity computing with agents and intelligent multimedia and databases. The application areas the author had started with are: portfolio management, forecasting, planning applications to OR, and micro-managing decision support with means-end analysis with agents. Multiagent business objects and Intelligent Multimedia applications to Management Science and IT are defined. Amongst the areas are agent computing, modern portfolios, heterogeneous computing, business objects, intelligent databases, intelligent objects, and decision science.

Chapter 9 is on multiagent planning, spatial computing, and practical aerospace design applications. Design techniques and a problem solving paradigm with ⟨object coobject⟩ pairs and multiagents are presented. A basis to spatial multiagent visual space navigation is presented. AI world revision techniques for intelligent multimedia with the Morph Gentzen logic are presented in brief. The techniques make it easier to formalize this area of research in AI towards a science. A technique for design and implementation of robot supervisory systems for a spacecraft controller with multiagent AI defined on the visual filed is presented. Multiagent implementations are applied to design robot supervision systems with autonomous double vision computing. A particular formulation of robot supervision is presented based on the agent models of computation. Many communicating pairs of multiagent modules define a system. A preview to a new model-based design for automatic onboard spacecraft visual field prediction based on Morph Gentzen computing is stated. The Double Vision AI Computing paradigm's applications to autonomous spacecraft vision and VR navigation are presented.

Chapter 10 is on new KR techniques with predictive model diagrams, planning with incomplete diagrams, and the applications to computable models for relevant worlds. G-diagrams are applied to KR to define localized keying to models and a minimal efficient computable way to discover specific models from knowledge represented with database applications. G-diagrams are applied towards KR from planning with nondeterminism and planning with free proof trees to partial deduction with predictive diagrams presented.

A specific completeness theorem is proved on predictive diagrams for partial deduction applications. The application to proof abstraction is alluded to and presented further in our papers. Diagrammatic reasoning is defined in terms of inferences directed by the G-diagrams for models. Thus, a basis for AI computing with GF diagrams is presented with applications to robot planning. All chapters, except for the introduction, have assigned exercises for academic instructions.

Chapter 11 presents Business Intelligence, Content Exploration, and Intelligent Interfaces. Agent trees and intelligent visual databases provide a basis for designing intelligent multi-tier interfaces with agents and intelligent business objects with applications to intelligent WWW interfaces. Basic intelligent content management with multi-tier designs for interfaces are percented. The overview to game trees and splitting agents applications to enterprise resource planning (ERP) and modeling are presented. The field of automated learning and discovery has obvious financial and organizational memory applications. There are basic applications to data discovery techniques with intelligence multimedia databases. The game trees are applied to train and reach onto models via learning to discover models from data mining (DM). The computing model is based on a novel competitive learning with agent multiplayer game tree planning. The computing techniques, the Morph Gentzen deductive system and its models are applied towards an active multimedia database warehousing, model discovery, and customizing interface design. Intelligent visual computing paradigms are applied to define the multimedia computing paradigm and active databases. The Intelligent Multimedia paradigms can be applied to databases and query-processing applications to stocks.

Chapter 12 is on design techniques with software agents and Abstract *Intelligent Implementations*. Agent morphisms are defined and applied to preservation principles. The object level definitions for individual modules can be automatically programmed by source abstract syntax tree to target abstract syntax tree morphisms. AII techniques are applied to define an Ontology Preservation Principle for Heterogeneous KB Design and implementation.

AI agents and Mediators define the stages of conceptualization, design and implementation. Multiagent implementations are applied to software design techniques, which incorporates object level nondeterministic knowledge learning and knowledge representation developed in Objects, message passing actions, and implementing agents are defined by syntactic constructs, with agents appearing as functions. By defining specified agent activators, events and activity are computed for the AII agents. The proposed AII techniques provide a basis for an approach to automatic implementations from intelligent syntax trees. Mediators implementing objects and agents define interpretability. Intelligent fail-safe systems are formally defined.

Abstract Planning, Models, and Virtual Trees are presented next. The notion of a generalized diagram and the techniques of model building for AI world revision in earlier papers of the author is further developed to formulate Free Proof Tree plan computation, model-theoretic planning, and generic model Diagrams. G-Diagrams are presented and the application to algebraic plan computation is briefed. A sound free proof tree computation theorem and a Hilbert Model theorem for plan computation with free proof trees is presented. Ordinal computing for proposition rankings can be applied as a basis for truth maintenance and planning. The structural approach to the problem of planning and reasoning in artificial intelligence that has been put forth in our papers can be applied to AI problems in general. The techniques are based on the AI worlds defined by G-diagrams for models. Thus, a basis for AI computing with GF diagrams is presented with applications to robot planning. The model-based planning techniques are applied with our new Morph Gentzen logic as a basis to design multiagent visual planning.

The Final chapter is on Intelligent Cyberspace Interfaces and WAP Generic Computing. A new view to WWW interface computing with agents is presented in our project and briefed here with the specific technical areas enunciated. Interactions amongst heterogeneous computing resources are via objects, multiagent AI and agent intelligent languages. New applications to business with intelligent object languages are presented in brief. The models can be applied as a basis to authentication on cyberspace computing. A basis to agent computing at virtual sites and functions is presented to allow insulated virtual engagement and remote invocations. WAP VPN generic computing is presented as specific techniques to design docking. Cybersigatures applications to mobile computing and Agent J2ME are examined.

REFERENCES

1. A. Barr and E. A. Feigenbaum, "The Handbook of Artificial Intelligence," Vol. 1, Chap. III, pp. 143–222. Kaufmann, 1981.
2. R. J. Brachman and H. Laveseque, Eds., "Readings in Knowledge Representation." Morgan-Kaufmann Publishers, Los Altos, California, 1985.
3. R. J. Brachman, R. E. Fikes, and H. J. Levesque, "KRYPTON: A Functional Approach to Knowledge Representation" (Brachman and Levesque, Eds.). In Knowledge Representation. Morgan-Kaufmann, 1985.
4. R. M. Burtall and J. A. Goguen, "Putting Theories Together to Make Specifications." Invited Paper, IJCAI, 1977.
5. C. Carson, M. Thomas, S. Belongie, J. M. Hellerstein, and J. Malik, "Blobworld: A System for Region-Based Image Indexing and Retrieval," pp. 509–516. In Third International Conference on Visual Information Systems. Amsterdam, 1999.
6. Chris Bailey_kellogg and Feng Zhao, "Qualitative Spatial Reasoning: Extracting and Reasoning with Spatial Aggregates," Vol. 24. No. 4, AI Magazine, AAAI Press, 2004.
7. M. G. Christel, A. G. Hauptmann, H. D. Wactlar, and T. D. Ng, "Collage as Dynamic Summaries for News Video," In Proceedings of ACM Multimedia. Juan Les Pins, France, 2002.
8. M. G. Christel, M. A. Smith, C. R. Taylor, and D. B. Winkler, "Evolving Video Skims into Useful Multimedia Abstractions." In Proceedings of ACM CHI'98 Conference on Human Factors in Computing Systems, 1998.
9. R. Fikes and T. Kehler, *CACM* 28, 904 (1985).
10. R. E. Fikes and N. J. Nilsson, *Artificial Intelligence* 2, 189 (1971).
11. A. Fujii, K. Itou, and T. Ishikawa, "A System for On-demand Video Lectures." In AAAI 2003 Symposium on Intelligent Multimedia Knowledge Management, Stanford, U.S.A., 2003.
12. G. Gentzen, *Math. Ann.* 119, 140 (1943).
13. M. Genesereth and N. J. Nilsson, "Logical Foundations of Artificial Intelligence." Morgan-Kaufmann, 1987.
14. A. G. Hauptmann and M. J. Witbrock, "Informedia: News-on-Demand Multimedia Information Aquistion and Retrieval" (M. T. Maybury, Ed.), pp. 215–239. In Intelligent Multimedia Information Retrieval, AAAI/MIT Press, 1997.

15. Jennifer Kahn, "It's Alive." Wired Magazine, 2002.
16. G. J. F. Jones and R. J. Edens, "Automated Management of Audio-Visual Presentations for Online Learning Applications." In AAAI 2003 Symposium on Intelligent Multimedia Knowledge Management, Stanford, 2003.
17. J. Keyes, "The Ultimate Multimedia Handbook." McGraw Hill, 1996.
18. S. Kleene, "Mathematical Logic," John Wiley and Sons, 1967.
19. Macromedia Interactive—Director Multimedia Studio. Peachpit Press, Berkeley, CA, 1996.
20. J. McCarthy, "Programs with Commonsense" (Brachman and H. Levesque, Eds.). In Semantic Information Processing. Readings in Knowledge Representation, Morgan-Kaufmannn, Los Altos, CA, 1985.
21. K. McKeown and D. Radev, "Generating Summaries of Multiple News Articles," pp. 74–82. In Proceedings of the 18th Annual International ACM SIGIR Conference on Research and Development in Information Retrieval, ACM, Seattle, 1995.
22. M. Mitra and B. B. Chaudhuri, *Information Retrieval* 2, 141 (2000).
23. N. J. Nilsson, "Searching, Problem Solving, and Game-Playing Trees for Minimal Cost Solutions" (A. J. Morell, Ed.), Vol. 2, pp. 1556–1562. North-Holland, Amsterdam, 1969.
24. N. J. Nilsson, "Problem Solving Methods in Artificial Intelligence." McGraw Hill, New York, 1971.
25. C. F. Nourani, "Equational Intensity, Initial Models, and AI Reasoning," Technical Report: A Conceptual Overview, in Proc. Sixth European Conference in Artificial Intelligence. North-Holland, 1984.
26. C. F. Nourani, "Visual Computational Linguistics—The Preliminary Definition." 1995.
27. C. F. Nourani, "Multiagent Chess Games," AAAI Chess Track, Providence, RI, 1996a.
28. C. F. Nourani, "Multiagent AI Design a New Emerging Software Engineering Trend," Vol. 12, No. 1, pp. 37–42. Engineering Applications AI—EAAI. Elsevier, 1996b.
29. C. F. Nourani, "MIM Logik." Bulletins Symbolic Logic, ASL, Prague, 1997a.
30. C. F. Nourani, "Versatile Abstract Syntax Meta-Contextual Logic and VR Computing 35th Lingustische Kolloquium." Innsbruck, Austria, 1997b.
31. C. F. Nourani, "Intelligent Trees, Thought Models, and Intelligent Discovery," Model-Based Reasoning in Scientific Discovery MBR'98. Pavia, Italy, 1998a.
32. C. F. Nourani, "The Agent Process Model for Intelligent Object Level Computing," Proceedings IFL 1999. Utrecht, 1998b.
33. C. F. Nourani, "MIM Logik." Summer Logic Colloquium, Prague, 1998c.
34. C. F. Nourani, "A Parallel Intelligent Object Module Language." Wkshp Parallel Functional Languages, London, 1998d.
35. C. F. Nourani, "Intelligent Multimedia—New Computing Techniques and Its Applications" (Ch. Freytag and V. Wolfengagen, Eds.). Proceedings of 1st International Workshop on Computer Science and Information Technologies, MEPhI Publishing, Moscow, Russia, 1999.
36. C. F. Nourani, "Autonomous Double Vision Spacecraft." AA99 Autonomous Control Track, Seattle, 1999a.
37. C. F. Nourani, "Aesthetics, CL, and Models Friedrich-Schiller-Universität Jena" Institut für Germanistische Sprachwissenschaft, Institut für Anglistik/Amerikanistik, "The 37th Linguistic Colloquium," Germany, 2002.
38. C. F. Nourani and K. J. Lieberherr, "Data Types, Direct Implementations, and Knowledge Representation," Vol. II, pp. 233–243. Proc. 19th HICSS. Honolulu, Hawaii, 1986.
39. R. W. Picard, "Affective Computing for HCI," Proceedings of HCI. Munich, Germany, 1999a.
40. D. Prawitz, "Natural Deduction: A Proof Theoretic Study." Stokhlom, Almqvist and Wiksell, 1965.
41. Y. Rui, A. Gupta, and A. Acero, "Automatically Extracting Highlights for TV Baseball Programs." In Proceedings of the Eighth ACM International Conference on Multimedia, pp. 105–115. ACM, Marina del Rey, CA, 2000.
42. A. Salway, M. Graham, E. Tomadaki, and Y. Xu, "Linking Video and Text via Representations of Narrative" University of Surrey. In AAAI 2003 Symposium on Intelligent Multimedia Knowledge Management, Stanford, 2003.
43. R. Wayraugh, *Artificial Intelligence* 13, 133 (1980).
44. D. Wilkins, "Domain Independent Planning: Representation and Plan Generation Artificial Intelligence."

CHAPTER 2

Creative Multimedia Techniques

CONTENTS

1. INTRODUCTION

This chapter is an overview to practical multimedia programming based on the available techniques and tools. Creative techniques are examined and the applications to animation, movies, video and virtual reality (VR) are presented. What is here does not apply AI yet. It is only the basic programming areas. The following chapters carry on Intelligent Multimedia with AI. Starting with creative techniques, some basic multimedia film and video programming and recording is presented with Adobe and the Director. Virtual reality and 3-D imaging is introduced. Next, morphing and its motion picture applications at real studios are outlined. The chapter concludes with some practical animation specifics.

2. CREATIVE TECHNIQUES

There are programming tools Adobe and Director, for example, which allow you to combine and edit source on clips and create movies. Digitized movies can be created by combining video, audio, and animation allowing you to play digitized movies on a PC. The programming steps on Adobe (1994) are as follows. The programmer starts by

1. Creating a new project.
2. Import clips into the project windows.
3. Assemble clips in the construction windows.
4. Preview the movies.
5. Save the project.

ISBN: 1-58883-037-3/$35.00

Intelligent Multimedia Computing Science
By Cyrus F. Nourani
Pages: 13–21

6. Compile a movie using the make movie command.
7. Play the movies in the clip window.

The video terminology is:

Clip—The source material for a movie, a stillimage, and an audio file.
In Adobe a clip is a pointer to a file stored on disk.
Frame—The basic unit of information in television, video, or digitized movies.
Timebase—The timebase for video in North America is 30 frames/second fps.
Many countries have 25 fps time base. Film plays at 24 fps. The duration of a clip is in terms of Hour:Minute:Seconds: frames.

Capturing The process if digitizing analog source material (video or audio), and then recording the images and sounds directly to your PC is in Adobe movie or audio capture commands. The capture can be at real-time.

An example is putting clips in the construction window. The construction window is used to arrange clips into a movie. The construction window contains over 100 tracts for clips, including the T track, for transition between clips, and the SI track for superimposed video clips. Tracks are indicated in the vertical bar at the right window: note that the lower tracks are the audio clips. At the top of the construction window is a line rules that indicate where each clip starts and ends. When the clips are put together in the construction window a compile time movie can be created. Apple's Quicktime is software that can then be applied to combine the video, audio, and animation to play digitized movie on a Macintosh.

The Director, which is where we review the applications to movies and animation in Section 2.4, has a sophisticated interface. To have a glimpse as to how the Director can be applied to multimedia programming, a sample Director Screen for Windows is depicted as follows.

Member ⟨file edit, view, Insert, Modify, Control, Xtras, Window, help⟩

Toolbar	
Score Window	A Frame-frame record of your Director Movie
Cells	The score is composed of cells, the storage unit that holds all the information about your movie. The columns of the cells are organized into frames. The rows make up the score channels.
Cast Window	The storage place for cast members in the Director. A cast window is organized into rows and columns of small windows, each containing a unique cast member.
A Cast Member	A multimedia component, e.g., bitmaps, sound
Control Palette	Processes VCR type control over play backs of your movie, rewind, play back. etc.
The Stage	Background at which your movie plays.
Paint Window	built-in paint programs

Some standard 3D imaging techniques are applied by systems designed since 1998, for example 3D Studio MAX, is morphing where the shape of one object is changed over time.

3. VIRTUAL REALITY

Virtual reality is a technique whereby applying computers 3D images are created with which one can navigate and interact. Virtual is being something in effect but not in actual form or name. It is a revision of reality without requiring the physical to build reality. It is not the same as what is known Imagineering as in the special effects brought on at Disney Worlds. It is not only the stereoscope images that bring on the VR. The HMD-Head Mounted Display immerses the user in the virtual world, immersion increasing the sensation. Computer games are not VR; neither are automatic generated special effects since there is no live stage awareness of the virtual characters.

Cyberspace has introduced new multimedia sensations whereby a mindspace travel might be a virtual perception while its technical basis is not what is exact VR in its present stage.

Morse (1998) is a nice start. The VR benefits are 3D visualization, navigation and immersive VR. Desktop VR is the ability to navigate without physically moving. It opens an unusual range of applications to VR. It allows you to visualize what would be visible at all rooms in a building. Hence a person can be familiarized with what the surroundings for an office building or laboratory feels before it is physically built. Interactions with virtual objects, Immersive VR, simulations, virtual environments are some benefits.

What is applied to get to VR: 3D Computer Graphics; modeling objects, data gloves, a see through HMD designed at MIT, 1982. Since then, there has been computer designed VR hardware and software. Dynamic objects are applied with collision detection avoidance at VR. It is sometimes achieved by virtual boxes as shields. Perspective views, 3D clippings, and stereo vision are created with VR. In the following chapters, we see how stereo vision is applied with intelligent multimedia. Images are rendered in form of pixels and displayed open on screens or HMD. Light sources area added as an extra realism degree. Rendering algorithms are applied, for example Painter's or scattering to render pictures. Painter's algorithm is similar to the way a painter paints a picture. Scattering algorithms are applied as when TV images are transmitted in from of horizontal rashers or scan times. The image is rendered raster by raster in the algorithm. Field depth is introduced by extra axes and computed by specific algorithms. Texture mapping is applied to bring forth realism on reflections, shadows, and spotlight.

A photograph might be scanned into the computer and superimposed onto the computer image. Environmental mappings simulate the effect of polished surfaces that reflect the surroundings. Shadows are brought on by applying the geometric distance between light sources and virtual objects. Radiosity is a rendering technique that creates photosensitive images. For example, ray tracing is an excellent technique for realizing shadows, reflections, refraction, etc. Fog is brought on by fades on objects. Transparency: Since light travels slower in materials such as glass and water, its direction changes and gives rise to refraction. The rendering sequence is important when there are transparent objects. If a transparent object is rendered followed by an opaque object, no information remains on the opaque object visibility degree when the transparent object is sampled. A popular science brief on the subject is Vince (1998). Other graphics techniques applied are: Soft objects—applies boundary representation and surface geometry with equations; particle systems—models large number of points; and mathematical curvature techniques. There are areas being called Techno Fetishism and sexy robots written for the virtual age.

4. INTRODUCTION TO MOTION PICTURES

4.1. Basic Morphs and Motion Pictures

A goal on the project is motion picture design and editing. In the present section and what has been stated above we have only gotten our feet wet. The motion picture enterprise might apply intelligent multimedia yet to be introduced. We had put forth the basis to Intelligent Multimedia morphing and its applications to motion pictures since 1996. Some static basis, i.e., without introducing AI, the area has been laid out thus far above to be augmented by basic morphing and 3D imaging. A thorough treatment to morphing is on the following chapters. Morphing has been applied by Lucas films mythic fantasy Willow design by ILM

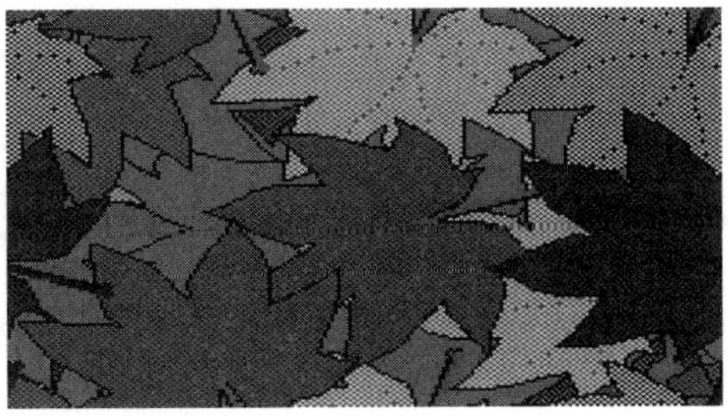

Figure 1. Morphed leaves.

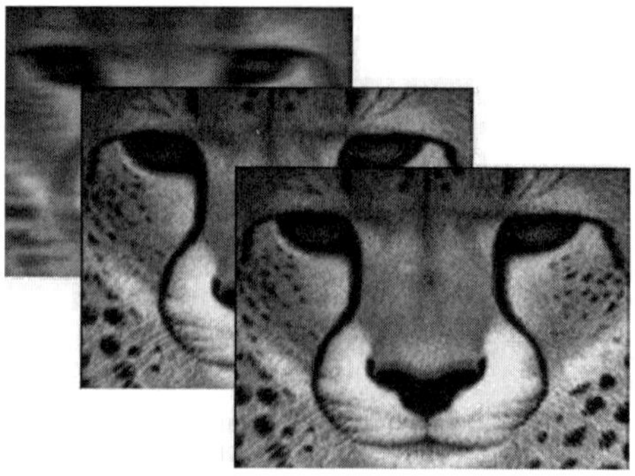

Figure 2. Image morphing indicating distance.

and Kodak. Willow introduced morphing as an image processing technique through which one image could be progressively transformed to another.

Basic programming with Morph 3-D imaging is stated here. Its mathematical logic basis is introduced with intelligent multimedia in the following chapters based on the author's technical publications presented in brief. To start with morphing is what is known as Shape Animation. Modeling the morph image is by applying a library of image poses on spline or polygonal models. Mathematical morphing techniques and applications were put forth by the authors since 1996. The chapters to come present the techniques and the application areas further.

A compound object creates and selects morphs. Multi-Image Morphing is a technique where a group of objects are image models. For example, as many as 100 morph images with different shapes are put into what is called a channel in Platinum Pictures augment to the Studio with a channel control menu. The morphs between any number of shapes and ratio are stored. The Morph can be limited by selecting vertices on individual objects. Another techniques is called Weighted Channel Morphing where with selected image morphs, channel spinners can be selected to represent a channel.

The channels might be combined a multiplexed based on the spinners to get new combinations. Morphings defined in the section are quite sophisticated bases to what we might call standard or static morphing without applying Intelligent Multimedia yet. Virtual static morphing might be defined to create lightening storm effect. It requires animating a light stage between two states with or without lightning. Animating lighting depends on how you select frames/second and the contrast between foreground and background lightening control. Basic mathematical morphing techniques were introduced by the author in 1996 (see Chapters 3 and 4). Windows logo is a composition morphing example. The following

Figure 3. Morph composition.

example, courtesy Christian Wild, A + Morph, Berlin, is how we might perceive composition morphing applied to 3-D designs with the techniques the presented to choose a specific design.

4.2. Animation Techniques

Applying the Director Windows we have a start on programming movies and animation. The way it is done at the Director is by (a) changing the position of sprites on the stage over successive frames (translation); (b) By arranging an alternating sequence over successive frames. Sprites are one instance is a graphical cast member in one specific cell in an animation channel (a graphical cast member types in channel bitmaps, PICT, shape, text, filed, buttoned, digital video).

(a) is supported by Step Recording; In-Betweeining, and real-time Recording (RTR).
(b) is supported by Cast to time, space-to Time, and the Paste Relative commands.

Step Recording is whereby you arrange each frame of animation by hand. It is the most labor-intensive method for the same reason. First the programmer arranges the sprites in a given frame, then records the frame; rearrange the sprites slightly and record the next frames; rearrange, record, and so on. The process of creating a series of frames all of which feature small differences in size and position of the frames makes animation possible. Step recording is not the easiest way to animation. A sprite is created as soon as you drag a graphical cast member from a cast window onto the stage, on into the score. Its function is that you can change its properties, size, blend percentage and which ink effect it uses; and the marginal cast member it is based on will not change. It allows you to vary based on the source cast member. The Director's Cast to time Command allows the programmer to place cast members into sequential frames. It is a way to create animation by moving a series of cast members from a cast Window into sequential frames in the Director's score. The positions of sprites on stage have to be arranged. In-Between Animation is the most practical animation because in-between features create the frames automatically. All the programmer has to do is to specify the start and end. The director generates the rest. The In-between command allows you to animate a sprite along a straight line.

Real-time Recording is the most direct way to animate with objects in the Director. With RTR a single cast member or a file loop cast member can be moved around the stage. The Director records its movement rate to the score automatically. The Director records its movement rate to the score automatically. Film loop is an animated cast member. It is composed of frame to frame animated sequence, for example a cyclist on its way, or a plane travelling the sky. What is nice is that it is a single cast member.

4.3. Practical Level Knowledge Representation (KR) and Morphs

Let us start with what is known as Full Body Morphing, a method that will allow you to make morphs that span over more than one part of the body (hip/abdo/chest combo). Applying modeling in general, and graphic intensive techniques, the method is in four stages:

1. Preparing the morph targets in with a model composing tool.
2. Designing the Morphs.
3. Converting them back to a user friendly interface to the composer tool.
4. Applying the morphs to the Poser figure.

Figure 4. In-betweening.

The above techniques have a foundation in the chapters in the present book and in the author's publications since 1995. What becomes blurry is how the menu driven design's computations have to do with the scientific basis. The close to reality analogy is to compare a science for object computation and the corresponding deductive systems to a menu driven interface to design a specific morph instance. The foundations are necessary to know what is being computed, what are the scientific properties of morph composition, and scientifically managing the underlying computations and compositions with a deductive system. On the practical side a browse had found sites were morphing is applied. For example, there are groups starting to apply morphs to design, A+Morph and SuperDesign. Germany, latter presenting sequenced morphs. There are the specifics on the practical designs on a typical system as an artistic tool not applying the scientific basis in our projects in a direct manner. To illustrate what goes on when a designer gets close to manually crafting morphs, the sites to browse to are tutorial.de and morphworld.com.

The former is an academic group, and the latter appears to be a specific practical application.

Example knowledge representation on specific frames is:

Displacementmaps	Pixels	3DMotionblurswitch
Raytracing		Shutter switch
Castshadows		
Filtering		Focaldistance

ImageOutputSettting

Frame size	Width-height-resolution	Frame-rate and count

Specifics:
Object Name:
Number of Vertices:
Min X = , Max X =
Min Y = , Max Y =
Min Z = , Max Z =
Number of Texture Vertices:
Number of Normals:
Number of Faces: 17238

4.4. Flash MX

Flash MX improves the ActionScript robots programming language and enhanced visual designer with animator. Flash has further an animation program for the Web and simple application development. A Flash designer might apply XML creating custom objects in addition to using motion tweens. Macromedia Flash MX goes further in introducing other tools to the Flash world through their ColdFusion and server products. Here are the areas where FLASH MX succeeds and where it struggles. The User Interface is an area where there is success. The controls are intuitive. The panels in Flash are the most essential controls for editing what's going on in the Stage and Timeline, or for authoring ActionScript.

The panels are collapsing panels that you can combine and arrange to your liking. The panel design allows you to dock several panels as a single unit and open as many as you can fit in your workspace. A single click expands the selected panel into full view and simultaneously hides other panels to save room. You can dock library panels with other panels or with each other. If you open multiple Flash documents, their libraries will be docked by default, making it easy to ascertain which library belongs to which open document. In the Windows version of Flash MX, you can dock everything to your workspace: the tools, panels, property inspector, and your document. The panel even includes Export, Import, and Print features. ActionScript author would find it quicker to type it by hand, but less-experienced users will probably be totally stumped. For some reason, this simple function, like others, is buried three menu levels deep. You can tinker with these menus by editing "ActionsPanel.xml."

There are far fewer panels. There are no panels on: Instance, Frame, Effect, Sound, Stroke, Fill, Text, Paragraph, and Character. If you use Dreamweaver 4, you may be familiar with this panel. It's a context-sensitive tool that most developers place at the bottom of their work space. It's many panels in one, and the options change depending on what you have selected. When you select text, text-related options are there. When you then select a frame, frame and tweening options take over. Select an object, and the Flash Property Inspector changes to reflect whatever object or symbol you have selected. It even presents movie parameters when nothing is selected. The result is that there are far fewer panels to hunt through. The Flash MX Timeline gives the needed attention to the frame selecting and manipulation function may seem subtle at first, but frame editing is much more intuitive in the latest release.

For example, double-clicking on a frame will select a range of frames, not open the Actions panel. Added layer folders help you organize complex projects. Contextual menus (right-click for Windows, Ctrl-click for Macintosh) now offer more relevant options. Images, sound, or other asset can be directly imported to the library. Dragging an object from the Stage to the library converts it to a symbol. When users create a new symbol, some want movie clip behavior and others want graphic behavior. The last behavior that a user selects stays. The video handling features in Flash MX are as follows. Flash MX supports video much like other assets. It includes the Sorenson Spark codec, which allows you to embed video directly into a SWF file. You can apply transformations to video much like other objects in Flash. However, your audience will need the Flash 6 player to view the movies. The following is a Flash example is our specific depiction design with Ina at ProMarkting and Design, www.prodesign-news.de, Berlin.

In Flash MX, the new loadMovie() function can also load JPEG files at run time via a URL. The new loadSound() function can likewise load an MP3 file. The movie's download is kept to a minimum and since the loaded files are not compiled and packaged with the Flash movie, it allows easier last minute changes to image files. The Free Transform tool adds more control over entire selections. The new Edit Envelope option for this tool allows more flexible distortionOne of Flash's big deficiencies is that users can't use their back

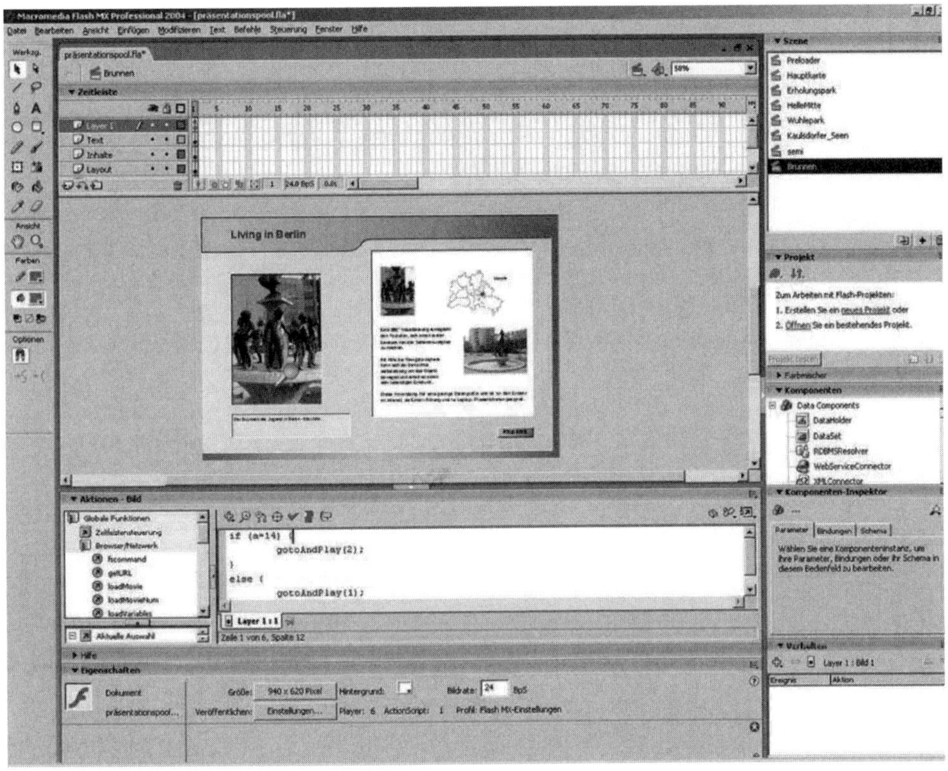

Figure 5. A flash window.

buttons or bookmarks within a Flash movie. The anchors in Flash MX allow you to assign this functionality where you want; however it doesn't work for Macintosh browsers. Many programs have design templates now, and Flash MX is no exception. Flash ships with several templates to help you build specialized content handled devices, such as Flash-enabled hand-helds like Casio's Pocket PC and the next generation of mobile phones. The ActionScript in Flash 5 allows developers and designers to create Flash-based games and applications. There is fast XML parsing, several new objects, and improvements to its scripting capabilities. Buttons are More Like Movie Clips (and Vice Versa). With Flash MX, you can name a button instance so that it can be the target of an action, but only the button instance, not its Timeline, is accessible as an object. Flash MX allows you to add button behavior to movie clips.

In Flash MX, you can use ActionScript to draw something from scratch. The new "createEmptyMovieClip()" function allows you to create a movie clip where none existed before. There are several methods for drawing straight-edged or curved objects within any clip. Flash MX components are basically the same things as Smart Clips, except more integrated into the Flash program. The new Components panel is a library just for these components. Packaged with MX are several components for adding interface widgets to your movies, like scrollbars and check boxes. It allows users to standardize navigation widgets so that scrollbars and checkboxes will always look consistent.

Flash MX carries out communicating with database content via XML. Macromedia has made a big move into Web applications development with Allaire. Now, with the whole Allaire team on its side, Macromedia has the resources to build a robust application development suite. Flash MX was built with ColdFusion in mind. In the near future, expect ColdFusion MX to offer full support for Flash MX output, along with other dynamic Web content. Flash MX is also designed to be integrated with other development platforms like J2EE and NET. That is the key to a whole new realm for Flash developers. Flash can be leveraged as a user interface development tool for Web-based applications with ColdFusion or other servers supplying the backend. These servers will be able to create real Flash objects and build Flash content without actually using Flash. Flash MX adds several new event handlers and objects for sound, microphones, and digital cameras. Macromedia is also developing a new communication technology that will make it possible to bring streaming video and audio to Flash.

Macromedia is betting on a future Web where users will be doing a lot more than browsing online, for example, on reservation systems and other commerce-related applications. While these developments may sound exciting on paper, how will they contribute to the real World of Web development? There are specific tutorials on what is applied.

Flash MX comes fast on with Adobe's LiveMotion 2.0. LiveMotion, which also runs in OS X, publishes SWF files that are viewable in a Flash player, and it has a few nice exclusive features like better frame rate management and integration with layered files from other Adobe products like Photoshop and Illustrator. LiveMotion 2.0 also has advanced scripting capabilities similar to those found in Flash 5—but not yet in MX. LiveMotion does not have a user-base wide enough yet.

EXERCISES

1. Combine and edit source clips to and create a movie with either Adobe or Director.
2. Apply Director to create an animation.
3. Digitize the movie from its source clip onto Quicktime to play digitized movies on a PC.
4. What aspects of VR can be designed by the techniques available by the tools presented in the chapter.
5. Macromedia is betting on a future Web where users will be doing a lot more than browsing online, for example, on reservation systems and other commerce-related applications. While these developments may sound exciting on paper, how will they contribute to the real world of Web development? What are three examples.

REFERENCES

1. Adobe. Adobe Premiere, Adobe Press, 1994.
2. W. Buckland, "Warren Film Studies—Teach Yourself." Hodder and Stoughton Publishers, 1998.
3. M. Cotta Vaz and P. R. Dunnignan, "Industrial Light and Magic—Introduction to the Digital Realm." Del ReyBooks, NY, 1998.
4. Ivan Sutherland, "The Ultimate Display." Vol. 2, pp. 506–508. Proceedings, IFIP Congress, 1965.
5. Macromedia Director. Peach-Pit Press, Berkeley, 1997.
6. M. Morse, "TV, Media, Arts and Cyberspace." Virtulaities. Indiana University Press, 1998.
7. 3D Studio, MAX, "Advanced Modeling and Materials" (Philip Miller, Ed.). Indiana, 1997.
8. Vince John, "Virtual Reality," p. 163. Springer-Verlag, 2004.
9. William B. Sanders, "Macromedia Flash MX." Seybold Seminars Publications, Wiley, NY-Indiana, 2003.

CHAPTER 3

Artificial Intelligence and Intelligent Multimedia Object Paradigms

CONTENTS

1. INTRODUCTION

This chapter starts with basic AI introducing AI agent computing. Multimedia AI systems are presented necessitating aesthetics with world models. Multimedia context computing and context abstraction is introduced leading into visual object perception dynamics. Intelligent multimedia hybrid pictures and hybrid morphing is presented. Next, a high-level multimedia language is defined and its applications outlined. OOP and IOOP—Intelligent OOP are introduced with visual objects leading onto Virtual Visual computing with JAVA Agent computing with mobile visual objects concludes the chapter. The agent process object

ISBN: 1-58883-037-3/$35.00

Intelligent Multimedia Computing Science
By Cyrus F. Nourani
Pages: 23–35

paradigm is a computing model which above its basic applications to the ordinary modern computing techniques, from ActiveX windows to event-driven computation, applets, mobile Java, and design with Object Fusion, has the capability to be lifted to intelligent multimedia computation.

2. Artificial Intelligence (AI)

The interest of AI for the past twenty five years in concepts and theories of logic, computational linguistics, cognition, and vision was motivated by the urge to bridge the gap from logic, metaphysics, philosophy and linguistics to computational theories of AI and practical AI systems. Practical artificial intelligence, AI, is concerned with the concepts and practice of building intelligent systems. Intelligent Multimedia techniques and paradigms are being defined whereby AI is applied to all facets of the enterprise starting from the basic multimedia programming stated in Chapter 1.

There are new computing techniques, languages, and a new deductive system with models. A new computing area is defined by Artificial Intelligence principles for multimedia. It is not a reinvention of AI, however, higher degrees of computational expressiblity is added when pictorial multimedia knowledge sources are augmented with agents and a complementing deductive system. Such computability is possible with the multiagent intelligent multimedia (MIM) and Morph Gentzen logic introduced by our papers.

3. AGENTS

The term "agent" has been applied to refer to AI constructs that enable computation on behalf of an AI activity. It also refers to computations that take place in an autonomous and continuous fashion, while considered a high-level activity, in the sense that its definition is software and hardware implementation, independent (Nourani 1991a). For example, in mission planning (Genesereth-Nilsson 1987, Nourani 1991b) or space exploration, an agent might be assigned by a designed flight system (Nourani 1991a) to compute the next docking time and location, with a known orbiting spacecraft. Intelligent agents are software entities that assist people and act on their behalf.

Intelligent agents can automate the retrieval and processing of information. *Software agents* are specific agents designed by a language that carry out specified tasks and define software functionality. Most agents defined by our examples are software agents. In the space applications we have had, of course, hard are functionality is specified for chip agents. Hence there are computing models with which agents might be applied the software counterpart to transistors and microchips.

The example depicted by Figure 1 is from our double vision (Nourani 1995c) and spatial computing projects. The visual field is represented by visual objects connected with agents carrying information amongst objects about the field, and carried onto intelligent trees for computation. Intelligent trees compute the spatial field information with the diagram functions. The trees defined have function names corresponding to computing agents. The computing agent functions have a specified module defining their functionality.

4. A MULTIMEDIA AI SYSTEM

Intelligent Multimedia is not mere AI techniques for manipulating independent multimedia interfaces. It is an entire new computing paradigm with a computing logic and foundations

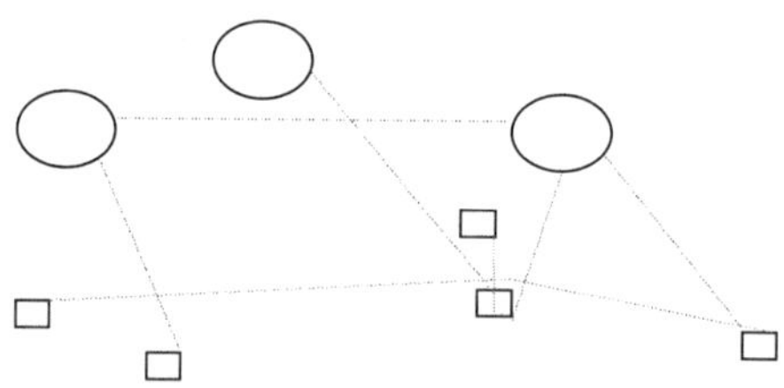

Figure 1. Visual agents and objects.

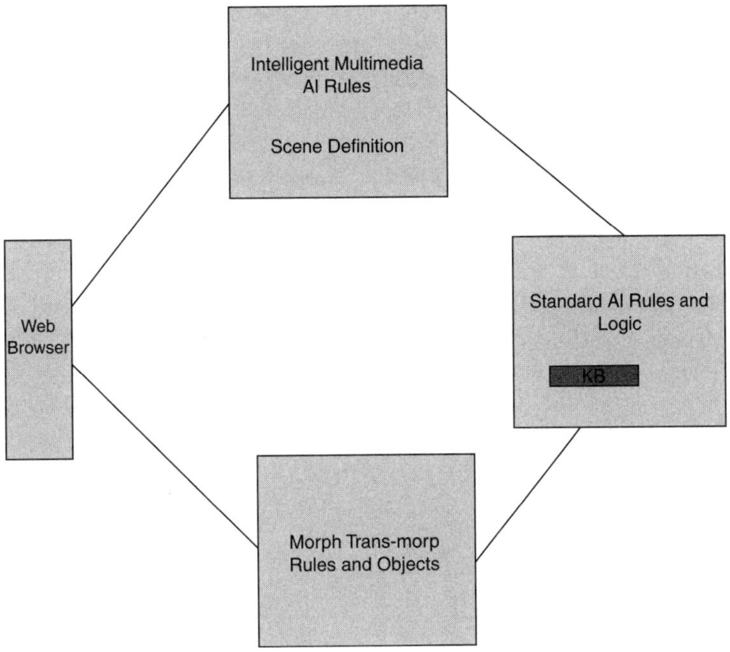

Figure 2. Intelligent multimedia WWW interfaces.

being developed. It is to be applied with multimedia interfaces to areas ranging from virtual reality tangible media and multimedia learning with interactive media, to television programming, telecommunications media, business and financial computing to multimedia commerce. It is further a development to the foundations to human thinking and learning Modern graphical interface techniques and explicit support for the user's problem-solving activities can be managed with IM.

The area for which the chapter starts on a foundation for is what is bound to be applied at dimensions and computing phenomena unimagined thus far, yet inevitable with the emerging technologies. The principles defined are practical artificial intelligence and its applications to multimedia. Multimedia AI systems are designed according to the new computing techniques defined. Multimedia Objects, rules and multimedia programming techniques are presented via a new language called IMPhora (Nourani 1996c). The concept of Hybrid Picture is the start to define intelligent multimedia objects. Trans-Morphing is the automatic hybrid picture transformation, is defined and illustrated by a multimedia language. A preliminary mathematical basis to an IM computing logic is presented in the subsequent chapters.

Multimedia Objects and Rules are presented and applied in programming. Hybrid Pictures are defined opening a new chapter to computing techniques. Trans-Morphing is presented as a dynamic computing principle applied to hybrid pictures and its computing importance is brought forth by way of new techniques and examples. It defines hybrid picture transformation. Intelligent Multimedia context defines the applications. Practical Multimedia Design is illustrated by pictorial examples. The application areas are based on advanced Artificial Intelligence available techniques. An intelligent multimedia interface to WWW might be depicted as follows from a 1996 project by the author.

5. AESTHETICS AND WORLD MODELS

According to Kant human knowledge is limited to appearances, whereas things in themselves are "noumena"—are thinkable but not actually knowable. Kant termed the doctrine Transcendental Idealism. Given the idealism is the possibility of synthesizing a priori knowledge to possible description and experience is easily explainable, since each object must necessarily conform to the conditions under which they can become objects for us. It assumes the human mind possesses such condition and demonstrating it is Transcendental Aesthetics. The computing import is since explored by our projects, e.g., Nourani (1998d).

To carry on with intelligent multimedia we have to be conscious as to what is the basic KR necessary to describe knowledge and its significance for perception. KR with G-diagrams for models (Nourani 1996d, e) and applications to define computable models and relevant world reasoning. G-diagrams are diagrams defined from a minimal set of function symbols that can inductively define a model. G-diagrams are applied to relevance reasoning by model- localized representations, and a minimal efficient computable way to represent relevant knowledge for localized AI worlds. We show how computable AI world knowledge is representable.

Basic application areas to start with as examples are designing predefined visual scenes with diagram composition and combination for scene dynamics. The second application area is based on AI planning (Fikes-Nilsson 1971, Nourani 1991). Reasoning and planning can be applied to define scene dynamics based on scene descriptions and compatibility relations. The project allows us to predict scene dynamics. We apply our recent Intelligent Language paradigm and intelligent visual computing paradigms to define the IM multiagent multimedia computing paradigm. By applying KR to define relevant worlds, personality parameters, combined with context compatibility and scene dynamics can be predicated.

5.1. Context

The role of context in KR and natural language systems, particularly in the process of reasoning (e.g., AIKR 1985) is related to diagram functions defining relevant world knowledge for a particular context. The relevant world functions can proliferate the axioms and the relevant sentences for reasoning for a context. A formal computable theory can be defined based on the functions defining computable models for a context (Nourani 1996d, 1997b). A preliminary overview to context abstraction and meta-contextual reasoning is presented from our papers (Nourani 1996d, 1997b).

Abstract computational linguistics with intelligent syntax, model theory and categories is presented in brief. Designated functions define agents, as in artificial intelligence agents, or represent languages with only abstract definition known at syntax. For example, a function Fi can be agent corresponding to a language Li. Li can in turn involve agent functions amongst its vocabulary. Thus context might be defined at Li. Generic diagrams for models are defined as yet a second order lift from context. The techniques to be presented have allowed us to define a computational linguistics and model theory for intelligent languages. Models for the languages are defined by our techniques in Nourani (1995b, 1996f). KR and its relation to context abstraction is defined in brief.

5.2. Visual Object Perception Dynamics

Visual dynamics based on general principles can be projected and the effects of scenes projected on viewers can be predicated. The ratings for TV programs can be predicted based on the relations amongst scene dynamics and viewer Dynamics. A problem-solving paradigm (Nilsson 1980) is presented in the Double Vision Computing paper (Nourani 1995c). The basic technique to be applied is viewing the televised scene combined with the scripts as many possible worlds. Agents at each world that compliment one another to portray a stage by cooperating. The AI techniques can be applied to define interactions amongst personality and view descriptions. The double vision computing paradigm with objects and agents might be depicted by the following figure.

The object co-object pairs[1] and agents solve problems on boards by cooperating agents from the pair *without splurges across the pairs*. The term splurge has a technical definition for object level computing presented in Nourani (1996f). Computing by agents might apply the same sort of cooperative problem solving (Nourani 1996). The IM paradigm can define multiagent computing with multimedia objects and carry on artificial intelligence computing on boards.

[1] The multiagent object-co-object computing design paradigm, invented in 1991, is the technical property of the the author's ventures. Commercial applications must be with proper credit and permission.

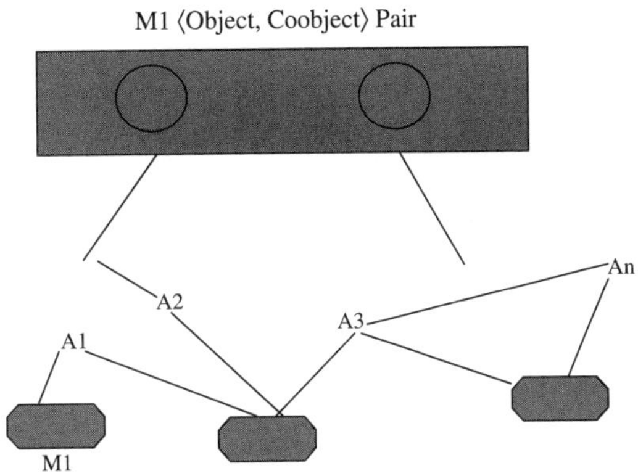

Figure 3. The object-coobject model.

The example below is illustrating where object programming is where a space age coffee-shop outlet scene is programmed.

Object := Coffee_Constellation
OPS := Serve_Coffee (Type, Table_no) | . . . . . .
Serve_Coffee (Spectacular_Brew, *n*) => Signal an available robot to fetch and serve
    (Spectacular_Brew, table *n*)
Exp := Serve_Coffee (Angelika, Table_no) | . . .
Serve_coffee(Angelika, Table_no) => if out_of_Angelika notify Table_no;
        Apply intelligent decision procedures to
        offer alternatives

In the above example OPS denotes operations, EXP denotes exceptions, and the last equation defines the exception action. APs are activities causing exceptional functions to be activated. Media examples are pauses and forgotten script lines by a personality being televised. In the example, there is a process (action) that is always checking the supply of Angelika coffee implementing the exception function. The circles are objects at a rectangle module. Ai's are agents and the dotted lines the message paths.

5.3. Hybrid IM Morphing

Since the IM visual objects are designed with agents the computation is via dynamic agent based morphs. A agent computation might cause a number of morph computations on pictures. The programming language IMPhora defines syntax for computing with programming constructs for Morphing, Hybrid Pictures and Trans-morphing. A Multimedia (Macro 1996) AI Systems program can be written in IMPhora. There are visual objects, rules and multimedia Programming. Most have seen hyperpictures on the WWW where there is pictorial view with text. There are underlined segments called hypertext. By selecting hypertext new windows are brought to view. *Hybrid Pictures* are the IM Hyperpictures which can be automatically transformed based on computing, images, or rules defining events, not only based on hypertext. Hybrid pictures are context and content sensitive hyperpictures.

Trans-morphing is a term the author invented to define automatic eventdriven or otherwise, hyperpicture transformation. Trans-morphing is the basic visual-computing event defined for hybrid multimedia computing. It is depicted by the following sequent figure. Figures combined with the Morph Gentzen rules, driven by specific inference rules might turn a rabbit out of a hat as depicted on Figure 4. The specific techniques and the mathematical rules are presented in the following Chapters.

6. A HIGH LEVEL MULTIMEDIA LANGUAGE AND PARADIGM

IMPhora is a high-level multimedia programming language designed (Nourani 1996b) to illustrate the structural morphings introduced by the intelligent multimedia computing paradigms. Its high level techniques allow the program designers to encapsulate agents and

Figure 4. The Morph Gentzen antecedents.

intelligent visual object modules while accommodating the parallel computation necessary for multimedia programming.

Agents are accommodated as processes.

IMPhora's Syntax

Small letters suffixed by integers are used to identify process.
Semicolon represents processes sequencing, for example, $p_1;p_2$
represents processes p_1 and p_2 running sequentially, p_1 followed by p_2
$\|$represents parallelism of either processes or modules, e.g., $p_1\|p_2$
$p_1 \rightarrow: \langle \text{msg} \rangle p_2$ represents the event p_1 sending a message to p_2.
In EBNF, the syntax of SLPX, is defined as follows.
⟨msg⟩ := ⟨data⟩|⟨request⟩|⟨(signal, pid)⟩|⟨(wait,pid)⟩
⟨status⟩|⟨error⟩|⟨halt⟩|⟨generics⟩|⟨⟩
⟨generics⟩ := ⟨special purpose msgs user supplied⟩
⟨⟩ := nil
⟨module⟩ := $\{p_1; p_2; \ldots; p_n\}\|\{q_1; q_2; \ldots; q_n\}\|\langle \text{module} \rangle I\langle\{p_i\}\rangle$
⟨Object⟩ := ⟨IMonitor⟩⟨module⟩|⟨Picture⟩|⟨HybridPicture⟩
Thus an IntMedia system can be defined by:
⟨IntMedia⟩ := ⟨module⟩$\|$
⟨HybridPicture⟩ :=
⟨Ops⟩ := ⟨ObjectOps⟩|⟨Morph⟩|⟨TransMorph⟩
IMPhora Conventions:
Small letters suffixed by integers are used to identify process.
Semicolon represents processes sequencing, for example, $p_1; p_2$
represents processes p_1 and p_2 running sequentially, p_1 followed by p_2.
$\|$represents parallelism of either processes or modules, e.g., $p_1\|p_2$.
$p_1 \rightarrow: \langle \text{msg} \rangle p_2$ represents the event p_1 sending a message to p_2.
In EBNF, the syntax of SLPX, is defined as follows.
⟨msg⟩ := ⟨data⟩|⟨request⟩|⟨(signal, pid)⟩|⟨(wait,pid)⟩
⟨status⟩|⟨error⟩|⟨ halt⟩|⟨generics⟩|⟨$\cdots$⟩
⟨generics⟩ := ⟨special purpose msgs user supplied⟩
⟨⟩ := nil
⟨module⟩ := $\{p_1; p_2; \ldots; p_n\}\|\{q_1; q_2; \ldots; q_n\}\|\langle \text{module} \rangle I\langle\{\text{pi}\}\rangle$
⟨Object⟩ := ⟨IMonitor⟩⟨module⟩|⟨⟩

Each of the message syntactic categories can be defined in very simple terms to suit the particular intended application areas. In addition, the specification language is extendible as

Figure 5. The Morph Gentzen bunny.

far as the type of communication it allows the processes and the nature of parallelism. The process bodies can be specified in terms of the actions they are intended to perform and the messages they send. Each process is a sequence of operations and messages, expressed as what is usually referred to as a process specification, in any specification language.

Monitors since Hoare are the known modular construct for parallel language synchronization and structuring. We define IM Monitors following new abstract monitors designed by the author Nourani (1993, 1998c).

$$\langle \text{IMS} \rangle := \langle \text{IMModule} \rangle \| g_1 \| g_2 \| g_3 \cdots \| g_n$$

where g's are guard processes to IMS.

6.1. Agent Object Oriented Programming

Objects are in the well-known sense of the word in object programming, abbreviated by OOP, for example. Objects consist of abstract data, perhaps encapsulation, and operations. Most recent programming techniques apply OOP in some form. Software engineering techniques with abstract data types have had OOP on their mind. Object programming started with Samltalk and C++ in the 80s. Its basis is, however, to a degree in the Norwegian Simula language projects at O. J. Dahl's group, and in programming with abstract data types, where the author had invested some years before OO, while at computing graduate school. It allows the programmer to design programs with structure based what the program is acting on not on what the program is doing.

The object classes and inheritence allows structure to be carried at abstract programs. Furthermore, encapsulation allows hiding unnecessary distracting information. Functions acting on data become part of data and can be encapsulated with structures. IOOP (Nourani 1995b) is a recent technique developed by the author combining AI and software agents with OOP. Since multimedia computing applies objects, and with IM, intelligent objects are introduced, where there are objects acted on by agents, we introduce visual JAVA as a possible basis to computing.

6.2. The Agent Process Model

New techniques presented and languages for object level programming with intelligent trees implemented by agent functions. It is shown how a two-level language paradigm and intelligent object level programming can handle what otherwise is a complicated computing phenomenon. The computing is with objects as situated automata, for which abstract syntax trees treatments and a computing theory merging with the current practice of programming theories are quite impossible. Intelligent syntax is put forth as the basis for automatic tree implementation techniques for object level programming by defining *OOPI*, OOP Intelligent, functions and Pull-up Abstractions as structural analogies to operation overloading. New linguistics constructs are defined for object level programming treating objects as abstract data types. The OOP defined by the present programming techniques have tree rewrite automatic implementations. Object computing systems are defined in terms of multiagent processes.

Computing with agents, mobile objects, and processes has presented computing with new powerful paradigms that must be put on a firm foundation before it can be comfortably applied by all. What our project addresses is a full treatment for the language syntax and semantics where there are mobile agents, processes, and objects with conventional tree computing techniques. There is a model theoretic soundness and completeness for the tree computing logic foundations to enhance the tree computing semantics. The present paper develops new techniques, language, and linguistics constructs for programming with objects implemented by agents. It is based on a theory of computing with trees on signatures that bear agent functions on trees. Thus, it is not sufficient to situate a programming theory for OOP in the object paradigm, especially for OOP defined for artificial intelligence. Additional mathematical abstractions, algebras, and logic are defined towards a computing theory for IOOP. Applications, as a theory of computing, to artificial intelligence, and object level programming, are stated in brief. These areas have been important goals for programming techniques. The concepts of intelligent syntax—thus languages are presented with applications to computing and OOP. Agent object algebraic tree rewriting was defined on intelligent trees by presenting the concepts and definition of algebraic tree TID and mutual tree information content within a forest.

6.3. Computing on Trees

In the projects referenced, the author has advocated programming techniques with intelligent trees and objects, where programming language automatic implementations and semantics can be defined from the abstract syntax trees and language constructs. The author has put forth the generalized diagram method to encode the model-theoretic semantics of a language from its abstract syntax. Language designs are presented with linguistics constructs that make it easier to identify G-diagram models and define automatic implementations from abstract syntax. Henkin defined the method of building models from syntax for first order logic. The computing enterprise requires more general techniques of model construction and extension, since it has to accommodate dynamically changing world descriptions and theories. The models to be defined are for complex computing phenomena, for which we define generalized diagrams.

The problem is that of defining object level programming with abstract syntax trees, abstract objects, which can be defined by G-diagrams and implemented by abstract Syntax trees. This paper is towards programming techniques that could provide a foundation for computing that is inevitable with the current and forthcoming AI programming techniques, supported by our theoretical developments. Technique for defining initial models by algebraic tree rewriting for OOP with intelligent syntax is to be developed from the authors techniques. It is shown how initial computing algebras can be defined by subtree replacement and tree rewriting as automatic OOP implementations. The authors doctoral research and publications over a decade ago pointed out the importance of the constructor signatures in computational characterization of initial models, and sufficient conditions for constructibility. Similar theories are being developed for OOP by defining intelligent languages and are presented in brief. However, the present theories have many new added theoretical turns and sophistication. The only common part is defining canonical models for tree rewriting theories.

6.4. Programming Linguistics and Intelligent Syntax

By an intelligent language we intend a language with syntactic constructs that allow function symbols and corresponding objects, such that the function symbols are implemented by computing agents. Sentential logic is the standard formal language applied when defining basic models. The language $\mathcal{L}$ is a set of sentence symbol closed by finite application of negation and conjunction to sentence symbols. Once quantifier logical symbols are added to the language, the language of first order logic can be defined. A Model $\mathcal{A}$ for $\mathcal{L}$ is a structure with a set A. There are structures defined for $\mathcal{L}$ such that for each constant symbol in the language there corresponds a constant in A. For each function symbol in the language there is a function defined on A; and for each relation symbol in the language there is a relation defined on A. For the algebraic theories we are defining for intelligent tree computing in the forthcoming sections the language is defined from signatures as in the logical language is the language of many-sorted equational logic. The signature defines the language $\mathcal{L}$ by specifying the function symbols' arities. The model is a structure defined on a many-sorted algebra consisting of S-indexed sets for S a set of sorts. By an intelligent language we intend a language with syntactic constructs that allow function symbols and corresponding objects, such that the function symbols are implemented by computing agents. A set of function symbols in the language, referred to by AF, is the set modeled in the computing world by AI Agents with across and/or over board capability. Thus the language $\mathcal{L}$ defined by the signature has designated function symbols called AF. The AF function symbols define signatures which have specific message paths defined for carrying context around an otherwise context free abstract syntax. Sets of function symbols in the language, referred to by AF, are agents with nontrivial capability. The boards, message passing actions, and implementing agents are defined by syntactic constructs, with agents appearing as functions. The computation is expressed by an abstract language that is capable of specifying modules, agents, and their communications. Here we have to define how the syntactic trees involving functions from the AFS are to be represented by algebraic tree rewriting on trees. This is the subject of the next section, where free intelligent trees are defined. The abstract syntax put forth to be implemented are expected at two levels. Level 1 is a language that only expresses the functionality of modules by names of objects and their message passing actions, for example by simple language for parallel xchange and processing (SLPX) (Nourani 1992). Level 2 defines the functions themselves. Amongst the new constructs introduced in SLPX are abstract monitors defined at syntax. The implementing agents, their corresponding objects and their message passing actions can also be presented by the two-level abstract syntax. From the practical standpoint, the models as individual programs can be specified with well-known specification languages. An important technical point is that the agents are represented by function names that appear on the free syntax trees of implementing trees. This formulation will prove to be an important technical progress. We can only present an overview in the present chapter and refer the reader to the mathematical basis chapters and author's papers for the theory. The trees defined by the present approach have function names corresponding to computing agents. The computing agent functions have a specified module defining their functionality. These definitions are applied at syntax tree implementation time. We define a function symbol f to be an **intelligent function** iff f is a member to the Agent Function Set. This is a free form definition that allows us to define tree algebras for intelligent computing theories. When we have specific computational linguistics applications, we can be precise as to what other properties the AFS is to have. For the present application area, the programming linguistics for OOP, we add an abstract computing relation to define OOP specific AFS.

6.5. Intelligent Object Level Programs

In the most recent language support for AI, programming, languages are pursued where objects are at play without much handle on the syntax trees. Objects appear on various semantic networks and are sort of situated automata that implement a computation by asynchronous methods. Our goal in the present project is to have language constructs that allow us to handle objects on abstract syntax trees and implement the mystical behavior of situated automata by agents.

It would be nice to view the problem from the standpoint of an example. The example of intelligent languages we have presented Nourani (1993) have $\langle O, A, R\rangle$ triples as control structures. O for objects, A's the activities and operations, and R the defining relations. The A's have operations that also consist of agent message passing. The functions in AFS are the agent functions capable of message passing. The O refers to the set of objects and R the relations defining the effect of A's on objects. Amongst the functions in AFS only some interact by message passing. What is worse, the functions could affect objects in ways that affect the information content of a tree. Thus the tree congruence definition is more complex for intelligent languages than those for ordinary syntax trees. Let us define tree information content for the present formulation.

DEFINITION. We say that a function f is a **string** function iff is no message passing or information exchange except onto the object that is at the range set for f, reading parameters visible at each object. Otherwise, f is said to be a *splurge* function. We refer to them by *string* and *splurge* functions when there is no ambiguity.

REMARK. Nullary functions are information strings.

Formal techniques with intelligent syntax, String and Splurge programming linguistics allow us to treat visibility in a precise way and to get theoretical results for MJOOP and OOP. These techniques and our formalization for AI computations lead to an exciting new programming theory and practice for MJOOP. The new programming techniques present the basis for programming with nondeterministic syntax.

DEFINITION OOPI. A function symbol f is *OOP intelligent* iff f is in AFS and it is decidable whether f is a string or splurge function.

DEFINITION. The tree intelligence degree, TID, is defined by induction on tree structures:

(0) the TID of a constant symbol function f is f;
(i) for a *string* function f, and tree $f(t_1, \ldots, t_n)$ the TID is defined by $\cup$TID $(t_i :: f)$, where $(t_i :: f)$ refers to a subtree of t_i visible to f;
(ii) for a **splurge** function f, TID is defined by $\cup$TID $(f : t_i)$, where $f : t_i$ is the updated tree t_i since a single message exchange to t_i from f.

Ordering on intelligence content is equivalent to subset ordering on TID sets. A rewrite rule on a set of trees with TID set S has information loss if the TID of the resulting rewritten trees is a subset of S. The methods of object level programming are implicit in the above definition. For example, the concept of a subtree being visible to a function refers to encapsulation methods. The theory usually emerges once language constructs are put forth for the programming paradigm. However, the reader might refer to the mathematical basis chapters for the theoretical developments for the present activity. Thus it is not necessary to believe the theory to accept the programming paradigms. Programming paradigms are often put forth without an *a priori* theoretical development.

6.6. The Objects KR Paradigm

Objects level programming has become quite prevalent at AI practice, where object appearing in various trees, almost by mystical means, by a situated automata behavior carry out computation. The computation, however, is quite transparent to our current theories of programming. It is the goal of the present project to shed some light on the AI and KR applications for OOP, as a programming practice, by providing a basis for a theory of computing with objects. Elsewhere a mathematical view to the Abstract object paradigm appears in Nourani (1998). Starting from Nourani-Lieberherr (1985) and the KRYPTON project (Brachman et al. 1985) might hint on how a functional KR paradigm might be carried out. With objects, we have a functional paradigm that deploys hierarchies and inheritance as a natural functional KR side effect augmented with agent trees. The author envisions that the method of intelligent syntax and computing on intelligent trees is to become a promising method of computing and in some sense inevitable programming paradigms for AI over the next decade. The interplay of our model theoretic method for computational linguistics with the method of intelligent syntax put forth here and the computability techniques could provide us with exciting expressive languages and techniques. The views of abstraction and object-level programming and the agent-computing basis are the important components of inter-play in the approach to KR with objects and agents. The intelligent syntax in that application area expresses computing agents and their message passing actions.

6.6.1. Beehived OOPI

Specific constructs are defined for object level programming with String and Splurge functions treating object visibility and messages. Logical properties are defined for the String and Splurge functions based on the decidability properties for the type theories. Thus an OOP abstract type is defined in terms of syntactic properties of the operations. Treating objects as abstract data types and a two level programming approach to OOP allows us to define Pull-up abstractions to treat incompatible objects. Pull-up is a structural definition comparable to overloading for functions. For example, the object Chair and Table might be pulled up to a Dinning Table object. The techniques define OOP type system with ⟨object-coobject⟩ pairs, thus a beehived computing systems is defined with the beehive edges which are ⟨object-coobject⟩ pairs. Agents defined on the hive's ⟨object-coobject⟩ pairs implement computation by string and splurge functions on the hives. The techniques open new areas of research on object level programming to pursue, and have put us at MJOOP, the Multiagent Junction For OOP.

6.7. Canonical Concurrency Calculus

We have seen thus far a specific canonical computing model for agent computing and mobile objects. Since the author's papers there is a new project at Sweden where a mathematical basis to a canonical calculus for concurrence is brought to light (Parrow et al. 1998).

Fusion calculus simplifies and extends the pi-calculus. Fusion contains actions akin to updating a shared state, and a scoping construct for bounding their effects. Therefore, it claims to easier computational models, e.g., the so-called strong reduction strategies in the lambda-calculus, involving reduction under abstraction. It is proved the calculus contains the pi-calculus as a subcalculus. The concurrency at our models are by canonical agent models and there are initial models for the computations. Object fusion calculus might be carried out with the agent model by an ambitious graduate student or colleague somewhere.

7. VIRTUAL VISUAL JAVA

A virtual machine is a conceptual computer mimicked inside a physical computer. Programs written for a virtual machine are written with a virtual instruction set. The para-instructions can be run wherever there is an emulator for the virtual machine. JAVA and its visual version J++ are programming languages for a virtual machine. A JAVA program is called an Applet. Applets are written to travel on the Internet and interact with CORBAs and onto IDLE-Microsoft's internet language. Applets can be complied into a virtual machine.

JAVA's syntax resembles C++ on the surface, but its goals and intentions are internet applications. Microsoft Visual J++ is an interactive development environment for JAVA. JAVA's J++ applets can carry image classes for details of working with graphics. The graphic object knows how to translate the image object bits into pixels. IMPhora programs might apply J++ to get physical visual objects carried around. Hence IMPhora can run on top of a J++ virtual environment. Nourani-Eshraglan (1998) is a basis to new multimedia pixel computing systems.

7.1. Mobile Visual Object Computing

There are Agent-based JAVA programming environments projects at IBM T. J. Watson Labs., and at the University of California Berkeley, which can be applied with IMPhora intelligent multimedia computing. The Berkeley EECS Java-To-Go is an experimental infrastructure that assists in the development and experimentation of mobile agents and agent-based applications for internet computing (interent computing: the set of applications that requires site-to-site computations. The main emphasis here is on a easy-to-setup environment that promotes quick experimentation on mobile agents.

Several such systems are reported at computing industry and at Stanford University. Java-To-Go is not intended to build artificial intelligence applications, nor should it be used to provide heavy distributed computation services. Java-To-Go is based on the Java programming language developed by Sun Microsystems. Java is widely known over Internet as the language for web programming. Java-To-Go provides an open (user modifiable) environment to transport actual Java implementation of the mobile agents across the network of agent servers. With Java's virtual machine, the same agent implementation can be run on all of the hosts. The industry versions might prove further polished.

EXERCISES

1. Design objects with agents with specific AI tasks to carry out a sequence consisting of objects, which depict an animated motion, based on messages from a motion director.
2. Carry out Exercise 1 with two figures communicating via agents in addition to the motion director.
3. Apply Chapter 2 specifics design languages you choose to Exercise 1.
4. Apply IMPhora style programming to carry on Exercise 2 with multiple actors or animated figures.
5. Apply IMPhora monitors to program virtual collision avoidance.

6. Apply visual JAVA and JAVA-to-go to design the intelligent objects in Exercises 1–5.
7. Apply the object KR paradigm to the design a basic active window interface to a store front as an EC system.
8. How do you apply agents with the KR in Problem 7 to deploy visual interfaces to specific items at the store inventory?

REFERENCES

1. S. Agamanolis and V. M. Bove, Jr., "Multi-Level Scripting for Responsive Multimedia." MIT Media Laboratory, IEEE MultiMedia, 1997.

2. AIKR, "Readings in Knowledge Representation" (R. J. Brachman and H. Laveseque, Eds.). Morgan-Kaufman Publishers, Los Altos, CA, 1985.

3. R. J. Brachman, R. E. Fikes, and H. J. Levesque, "KRYPTON: A Functional Approach to Knowledge Representation" (Brachman and Levesque, Eds.). In Readings in Knowledge Representation. Morgan-Kaufmann, 1985.

4. Byrne Morgan, "Virtual J++ 1.1 Unleashed," 2nd ed. Sams Net, Indiana, 1997.

5. R. E. Fikes and N. J. Nilsson, *Artificial Intelligence* 2, 189 (1971).

6. M. Genesereth and N. Nilsson, "Logical Foundations of Artificial Intelligence." Morgan-Kaufmann, 1987.

7. G. Gentzen, *Math. Ann.* 119, 140 (1943).

8. L. Henkin, *J. Symbolic Logic* 14 (1949).

9. J. Keyes, "The Ultimate Multimedia Handbook." McGraw Hill, 1996.

10. Macromedia Interactive—Director Multimedia Studio. Peachpit Press, Berkeley, CA, 1996.

11. N. J. Nilsson, "Principles of Artificial Intelligence." Morgan Kaufman, 1980.

12. C. F. Nourani, "Equational Intensity, Initial Models, and AI Reasoning." Technical Report: A Conceptual Overview, in Proc. Sixth European Conference in Artificial Intelligence. Pisa, North-Holland, Italy, 1984.

13. C. F. Nourani, "Planning and Plausible Reasoning in AI," pp. 150–157. Proc. Scandinavian Conference in AI, IOS Press, Roskilde, Denmark, 1991.

14. C. F. Nourani, "A Multi-Agent AI Approach to Fault Free and Fault Tolerent AI" (Douglas D. Dankell II, Ed.). ProjectMetaai, FLAIRS-93, Proc. Florida AI Symposium. Florida Research Society, St. Petersburg, Florida, 1993c.

15. C. F. Nourani, "Abstract Intelligent Tree Computing and A Decision Tree Rewrite Theory For OOP." Holmenkollen, Oslo, Norway, 1995.

16. C. F. Nourani, "Free Proof Trees and Model-Theoretic Planning." AISB, Sheffiled, 1995a.

17. C. F. Nourani, "Automatic Models from Syntax," Scandinavian Linguistics Conference. Oslo, Norway, 1995b.

18. C. F. Nourani, "The IOOP Project," Vol. 30, No. 2, pp. 56–54, SIGPLAN Notices, 1995b.

19. C. F. Nourani, "Double Vision Computing." IAS-4, Intelligent Autonomous Systems, Karlsruhe, Germany, 1995c.

20. C. F. Nourani, "IM A Super Multimedia Programming Language." Preliminary Report, 1996a.

21. C. F. Nourani, "KR, Diagrams For Models, and Computation," 3rd IMACS|IEEE International Multiconference on Circuits Systems. Communications and Computers, Athens, Greece, 1996b.

22. C. F. Nourani, "Multiagent AI Design a New Emerging Software Engineering Trend," Vol. 12, No. 1, pp. 37–42. Engineering Applications AI-EAAI, Elsevier, 1996c.

23. C. F. Nourani, "Intelligent Multimedia," Intelligence and Multimedia Applications. Poster Announcement. Monash University, Victoria, Australia, 1996d.

24. C. F. Nourani, "Slalom Tree Computing." AI Communications, The European AI Journal, IOS Press, Amsterdam, 1996e.

25. C. F. Nourani, "Intelligent Trees, Thought Models, and Intelligent Discovery." Model-based Reasoning in Scientific Discovery, Pavia, Italy, 1996f.

26. C. F. Nourani, "AII and Heterogeneous Designs." MAAMAW, University of Karlskrona/Ronneby, Ronneby, Sweden, 1997a.

27. C. F. Nourani, "MIM Logik." ASL, Prague, Bulletins Symbolic Logic, 1997b.

28. C. F. Nourani, "Versatile Abstract Syntax Meta-Contextual Logic and VR Computing 35th Lingustische Kolloquium." Innsbruck, Austria, 1997c.

29. C. F. Nourani, "Intelligent Languages" (A. Kelemenová, Ed.), pp. 281–287. A preliminary Syntactic Theory. In Proceedings of the Mathematical Foundations CS'98 Satellite workshop on Grammar systems. Silesian University, Opava, 1998a.

30. C. F. Nourani, "The Agent Process Model For Intelligent Object Level Computing." Proceedings IFL 1999. Utrecht, 1998b.

31. C. F. Nourani, "Intelligent Object Fusion." ACM Software Engineering Notes, 1998c.

32. C. F. Nourani, "A Parallel Intelligent Object Module Language." Wkshp Parallel Functional Languages, London, 1998d.

33. C. F. Nourani, "Intelligent Multimedia—New Computing Techniques and Its Applications" (Ch. Freytag and V. Wolfengagen, Eds.). CSIT'99, Proceedings of 1st International Workshop on Computer Science and Information Technologies. MEPhI Publishing, Moscow, Russia, 1999.

34. C. F. Nourani, "Aesthetics, CL, and Models Friedrich-Schiller-Universität Jena" Institut für Germanistische Sprachwissenschaft, Institut für Anglistik/Amerikanistik, The 37th Linguistic Colloquium, Germany, 2002.

35. C. F. Nourani and K. J. Lieberherr, "Data Types, Direct Implementations, and Knowledge Representation," Vol. II, pp. 233–243. Proc. 19th HICSS. Honolulu, Hawaii, 1986.
36. C. F. Nourani and K. EshragIan, "Autonomous Vision Computing and Smart Pixel Multimedia Chips," IV98. Stuttgart, Germany, 1998.
37. J. Parrow and B. Victor, "The Fusion Calculus: Expressiveness and Symmetry in Mobile Processes," pp. 176–185. In the Proceedings of LICS'98, IEEE Computer Society Press, 1998.
38. D. Prawitz, "Natural Deduction: A proof Theoretic Study." Stokhom, Almqvist and Wiksell, 1965.
39. R. P. Ten Dyke and J. C. Kunz, *IBM Systems Journal* 28, 3 (1989).
40. Vogel and Dully, "JAVA Programming with CORBA." John Wiley and Sons, Inc., New York, 1998.

CHAPTER 4

Basic Intelligent Multimedia Techniques

CONTENTS

1. INTRODUCTION

This chapter starts with visual context and objects and multiagent intelligent multimedia. Context abstraction and meta-contextual reasoning are introduced as a new field. Multi-agent visual multi-board planning is introduced as a basis to intelligent multimedia with applications to spatial computing. Practical goal satisfaction with practical example worlds applying morphings is examined. Visual model-based computing is introduced with the new techniques.

2. HYBRID PICTURES

Hybrid pictures were introduced in brief in Chapter 3. What distinguishes hybrid pictures in the intelligent multimedia context is the way agents correspond to the pictorial structure. Structures are designated by signatures similar to the way the key signature in music designates a structure. Signatures define the way the functions interact with the carriers for objects, scenes, and images. Signatures can designate color combinations. Multimedia objects are signatured structures on which morphings can be defined. A hue-morphism for example can change the hue on an image. Precise mathematical morphings can be defined and be computed with AI agents carrying on the intelligent morphing aspects.

Agents can be applied to change picture configurations and structure with AI computed morphing as responsive, proactive, and reactive live images. To start with motivations from artistic creativity let us compare some paintings. Monet's Cathedrals at Bright Sun, Sunlight,

ISBN: 1-58883-037-3/$35.00

Intelligent Multimedia Computing Science
By Cyrus F. Nourani
Pages: 37–45

and Fog, 1894 provides an example for specific creative morphings on hybrid pictures (Nourani 1998). The three cathedral paintings are based on a structure preserved by three morphings, each enunciating a certain hue morphism.

Another example is Monet's Sunrise 1872, where there is a morphing based on specific water whereupon a sunrise object and a boat are morphed. A third example appears in Pissaro's paintings. The three paintings are Early Frost 1873, Red Roof, 1877, and La Cote des Coeats 1877. There is an example of what might be a trans-morphing (see Sections 3–4). The building structure is trans-morphed with a structure preserving morphing and a hue morphing with some trees from the first two paintings to the third. Since the author wrote the above he has learned Adobe is as far as having Impressionistic effects built-in in its design techniques.

To apply the concepts with what we had started at Chapter 2 an IM Window parallel to the Director might be augmented with Intelligent Multimedia hyper picture as follows. The specific agent model is due to the author.

Hyper-Score Window	A Frame-frame record of your Hyper-pictures.
Agents	The score is composed of agents. Agents carry all the information about your movie. The columns of the cells are organized into frames. The rows make up the score channels.
Cast Window	The storage place for cast members. A cast window is organized into rows and columns of small windows, each containing a unique cast member.
A Cast Member	A multimedia component, hyper-picture agent code, bitmaps, sound
Control Palette	Processes VCR type control over play backs of your movie, rewind, play back, etc.
Hyper Stage	Background at which your hyper pictures play.
Paint Window	Built-in paint programs.

3. MORPHING AND TRANS-MORPHING

Trans-morphing is event driven picture transformation carried on by the hybrid picture agents. *Hybrid Pictures* are IM hyper-pictures, which can be automatically transformed based on computing, images, or rules defining events. Hybrid pictures are context and content sensitive hyper-pictures. Trans-morphing as automatic event-driven or otherwise, hyper-pictures transformations are a model for mind's, human or android, subconsience activities. Trans-morphing is the basic visual-computing event defined for hybrid multimedia computing by logical rules, we have called Morph Gentzen rules, on hybrid-picture figures. For example, $F_1\&F_2 \Rightarrow F_3$, where F_i's are hyper-picture figures.

Multimedia objects are viewed as syntactic objects defined by functions, to which the deductive system is applied. The syntax is specified with multimedia intelligent agents. We have defined *Intelligent Syntax*, *Intelligent Languages* and *Intelligent Tree computing* in Nourani (1996f, 1997d). By an intelligent language we intend a language with syntactic constructs and function symbols and corresponding objects, such that the function symbols are implemented by computing agents in the sense defined in Nourani (1996f), Genesereth-Nilsson (1987). Intelligent Tree computing (Nourani 1996f) has Soundness and Completeness theorems for its computational logic. Intelligent syntax carries cognitive computing.

The Double Vision Computing artificial intelligence paradigm is a multiagent computing paradigm where the agents cooperate for problem solving on boards. The world is viewed by dual pairs of possible worlds. Multimedia computing has since been defined by our projects. Morphing is applied to model subconsience imaging activities by applying the Morph Gentzen computing logic is a way to model discovery and a basis for a mind computing model (Nourani 1996e, 1997b) and here. The logic and the MIM Multimedia computing techniques have a computing logic counterpart. The basic principles are a mathematical logic where a Gentzen or natural deduction system is defined by taking multimedia objects coded by diagram functions.

Defining hybrid-pictures has opened a new chapter to computing techniques. The ideas are further applied to mind and cognition in Nourani (1996c, 1998a, b). Marvin Minsky's MIT development of the theory of human thinking and learning called the "Society of Mind," tries to explain how various phenomena of mind emerge from the interactions among many different kinds of highly evolved brain mechanisms. Reaching the goals to explain common sense, imagination, and reasoning by analogy paradigms in such projects can be accomplished by our multiagent multimedia basis to intelligence. Trans-Morphing is presented as a dynamic computing principle applied to hybrid pictures and its computing importance is brought forth by way of new techniques and examples. It is the metamathematics for the way multimedia reasoning is carried out on the human or robot minds.

The programming language IMPhora defines syntax for computing with programming constructs for Morphing, Hybrid Pictures and Trans-morphing. A Multimedia AI Systems program can be written in IM. IM includes Multimedia Objects and Rules and multimedia Programming. *Hybrid Pictures* are IM Hyper-pictures, which can be automatically transformed, based on computing, images, or rules defining events.

Hybrid pictures are context and content sensitive hyperpictures. Trans-morphing is a term the author invented to define automatic eventdriven or otherwise, hyper-picture transformation. Trans-morphing is the basic visual-computing event defined for hybrid multimedia computing. The IM Hybrid Multimedia Programming techniques have a computing logic counterpart. The basic principles are a mathematical logic where a Gentzen (1943) or natural deduction (Prawitz 1965) systems is defined by taking multimedia objects coded by diagram functions. By transmorphing hybrid picture's corresponding functions a new hybrid picture is deduced.

Multimedia objects are viewed as syntactic objects defined by functions, to which the deductive system is applied. Thus we define a *syntactic morphing* to be a technique by which multimedia objects and hybrid pictures are homomorphically mapped via their defining functions to a new hybrid picture. The deduction rules are a Gentzen system augmented by Morphing, and Trans-morphing. The logical language has function names for hybrid pictures.

The Morph Rule—An object defined by the functional n-tuple $\langle f_1, \ldots, f_n \rangle$, where f_i's are abstract object selectors, can be morphed to an object defined by the functional n-tuple $\langle h{:}\ f_1, \ldots, h{:}\ f_n \rangle$, where $h{:}f_i$ indicates the structure obtained by applying h to the structure f_i selects, provided h is a homorphism on the selected intelligent objects as abstract algebras (Nourani 1993a). Homomorphisms are operation preserving mappings amongst signatured structures (see chapters further on).

The *Trans-Morph Rules*—A set of rules whereby combining hybrid pictures $p_1, \ldots, p_n$ defines an Event $\{p_1, p_2, \ldots, p_n\}$ with a consequent hybrid picture p. Thus the combination is an impetus event. The important point is the picture p might not have anything obvious to do with $p_1, \ldots, p_n$. It is a hybrid-picture obtained by agent computation based on rules, schemes, the knowledge base and/or the active database. The deductive theory is a Gentzen system in which hybrid pictures are named by parameterized functions; augmented by the MIM morph and transmorph rules. The preliminary mathematics appeared in Nourani (1997).

A Morph rule example: The World has only circles, rectangles, and squares as objects. The only operation on the objects is composition $\circ$, where object a composed with object b, written $a \circ b$, modeled by a structure puts a above b. The 4-tuple in Figure 1, lower half,

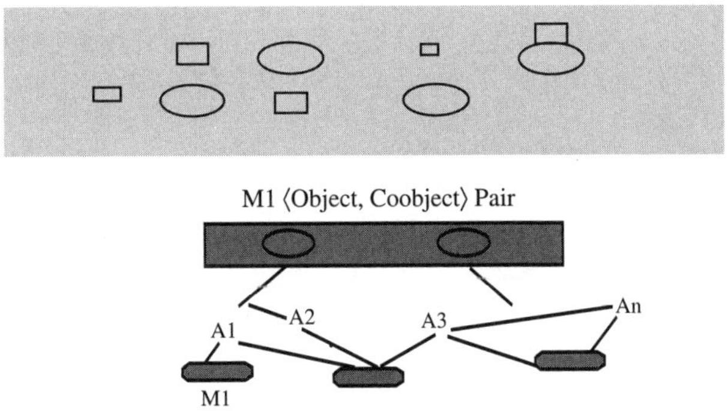

Figure 1. Morphing with composition.

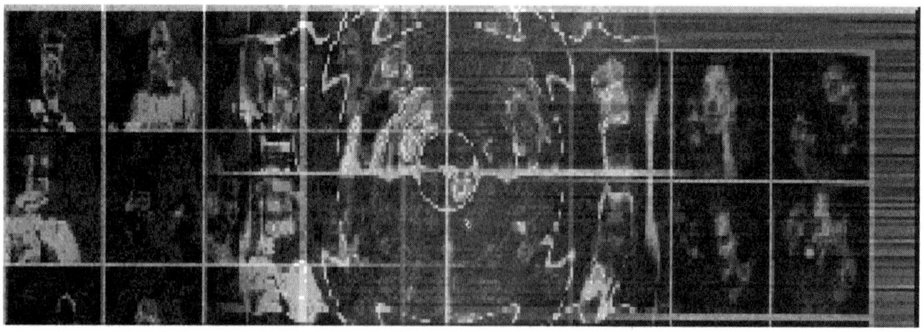

Figure 2. Morphing with composition, Section 3 Figure 2 liquid Morph Gentzen, Section 3.

is obtained by morphing $\langle f_1, f_2, f_3, f_3, f_4 \rangle$ where f_i's are functions selecting world objects, squares or circles, and $\langle f_1, f_2, f_3, f_4 \rangle \rightarrow \langle f_1, f_2, f_3, f_1 \circ f_2 \rangle$.

$$h(x) = x, \quad \text{for all } x \text{ a world member}$$

$$h(x \circ y) = h(x) \circ h(y)$$

Nontrivial structural morphings can be defined where the f_i's are selector functions on abstract structures with many carriers and operations, thus the resultant morph is with altered shapes. For example, obtaining a world composed with ovals and diamonds from a square-circle world. A trans-morphing example had been presented in Chapter 3. The above morphing techniques are rudiments to a virtual geometric computing science. The following is an example on liquid morph Gentzen form a project the author had written up a technical report on with Monika, where a specific example was designed to portray fluidity in applying Morph Gentzen. The project is preannounced on IMK, Sank Augustin, Germany, Netzspannung.org (Nourani-Flieshmann 2000).

4. INTELLIGENT LANGUAGES

By an intelligent language we intend a language with syntactic constructs that allow function symbols and corresponding objects, such that the function symbols are implemented by computing agents, e.g., Nourani (1997b). Sentential logic is the standard formal language applied when defining basic models. The language $\mathscr{L}$ is a set of sentence symbol closed by finite application of negation and conjunction to sentence symbols. Once quantifier logical symbols are added to the language, the language of first order logic can be defined.

A Model $\mathscr{A}$ for $\mathscr{L}$ is a structure with a set A (see Chapter 6). There are structures defined for $\mathscr{L}$ such that for each constant symbol in the language there corresponds a constant in A. For each function symbol in the language there is a function defined on $\mathscr{A}$; and for each relation symbol in the language there is a relation defined on A. For the algebraic theories we are defining for intelligent tree computing in the forthcoming sections the language is defined from signatures as in the logical language is the language of many-sorted equational logic.

The signature defines the language $\mathscr{L}$ by specifying the function symbols' arities. The model is a structure defined with many carrier algebras consisting of S-indexed sets for S a set corresponding to the carrier types for objects and classes. A set of function symbols in the language, referred to by AF, is the set modeled in the computing world by AI agents to which a computation scenario might assign missions across classes, modules, and objects computation and message passing missions.

Thus the language $\mathscr{L}$ defined by the signature has some designated function symbols called AF. There is a soundness and completeness theorem proved for the intelligent language logic (Nourani 1993a, 1998c). The theorems are to guarantee what is proved is sound and modeled by the structures for which the language is defined. The AF function symbols define signatures that have specific message paths defined for carrying context around an otherwise context free abstract syntax.

A set of function symbols in the language, referred to by AF, is the set of agents with nontrivial capability. The computation is expressed by an abstract language that is capable of specifying modules, agents, and their communications. We have put together the AI concepts with syntactic constructs that could run on the tree computing theories we are presenting in brief. We have to define how the syntactic trees involving functions from the AF are to be represented by algebraic tree rewriting on trees. This is the subject of the next section, where free intelligent trees are defined. An important technical point is that there are function names on trees for each agent.

5. PLANNING AND MORPHS WITH IM

A plan is a sequence of operations in a world model yielding terms that instantiate the truth of the goal formulas in the universe (Nourani 1995b). That is what goes on as far as the algebra of the planning world is concerned. It is a new view of planning prompted by our method of planning with diagrams for models and free Skolemized trees. It is a model-theoretic view. Proof-theoretically a plan (Fikes-Nilsson 1971), is the sequence of proof steps that yield the proof for the goal formula. The proof theoretic view is what the usual AI literature presents.

Since the basic IM computing visual objects are hybrid pictures we define new planning techniques with hybrid pictures. The IM planning does not only applies planning with agents, it applies Morph Gentzen rules to hybrid pictures to achieve plan goals. The hybrid pictures carryout responsive, proactive, and reactive planning, only initiated and directed by a planning system. An example IM planning mission (Nourani 1999b) is as follows.

Hybrid picture 1—Spacecraft A Navigation Window.
Agents: A1 Computes available docking times based on the visual field on the window.
A2 carries out docking sequence based on messages to Spacecraft B
Hybrid picture 2 Spacecraft B Navigation Window
Agents: B1 carries on course based on its visual field window
B2 Accepts and carries out docking maneuvers from external
hovering craft agents
Plan Goal Engage docking between A and B at appropriate A and B field windows.
Morph Gentzen computing can be applied to the hybrid pictures to satisfy a plan goal.
Thus morphing is applied with precise fluidity to plan computation.

6. PRACTICAL AI GOAL SATISFACTION

Practical AI systems are designed by modeling AI with facts, rules, goals, strategies, knowledge bases. Patterns, schemas, AI frames (Fikes-Kheler 1985) and viewpoints are the micro to aggregate glimpses onto the database and knowledge bases were masses of data and their relationships-representations, respectively, are stored. Schemas and frames are what might be defined with objects, the object classes, the object class inheritances, user-defined inheritance relations, and specific restrictions on the object, class, or frame slot types and behaviors.

A scheme might be
Intelligent Multimedia
IS-A Science
Fields AI, Multimedia, Mathematics, Art, Motion Picture
Member-of Academy of Arts and Sciences
Alternate
Intelligent Forecasting
IS-A Stock Forecasting Technique
Portfolios Stock, bonds, corporate assets
Member Management Science Techniques

Schemas allow brief descriptions on object surface properties with which high level inference and reasoning with incomplete knowledge can be carried out with facts and the defined relationships amongst objects. Relationships: Visual Objects A and B have mutual agent visual message correspondence. In the preceding section space example, the astronaut being between the spacecraft and a visible planet, is another example.

Looking for patterns is a way some practical AI is carried on with to recognize important features, situations, and applicable rules. From the proofs standpoint patterns are analogies to features as being leaves on computing trees. *Forward chaining* is a goal satisfaction technique, where inference rules are activated by data patterns, to sequentially get to a goal by apply the inference rules. The current pertinent rules are available at an *agenda store*.

Carried out rules modify the database. Backward chaining is an alternative based on opportunistic response to changing information. It starts with the goal and looks for available premises that might be satisfied to have gotten there. Goals are objects for which there is automatic goal generation of missing data at the goal by recursion backward chaining on the missing objects as sub-goals. Data unavailability implies search for new goal discovery.

Example: IF Goal: Intelligent Multimedia is applied to design motion pictures
THEN
Strategy: Morph Gentzen system is acquired by a major movie company Movie Foxes
AND Funding is obtained to complete the system
And MyVentures and Movie Foxes have signed a development agreement.

7. VIRTUAL MODEL-BASED COMPUTING

A proof of a first order sentence φ is a finite sequence of sentences starting with some known facts Δ, from the database for example, where φ is a member of the sequence (usually the last); and every sequence member is either a member of Δ, a logical axiom, or the result of applying the deduction rules, to the sentences appearing before on the sequence. For ordinary deductive systems the deduction rule is only modus ponens.

It is passing from two formulae p and $p \to q$, to q, where $\to$ is the logical implication symbol. p and $p \to q$ are called the premises, and q is called the conclusion. Proof trees are branch compositions on alternatives carried out to reach a conclusion from some premises. Alternative "bottom-up" proof trees are branch compositions starting from a goal to be proved to what antecedent and succedent it takes to get up to the goal, recurring on the antecedent and the succedent as goals.

Forward chaining and backward chaining is the AI term for it, respectively. Since planning applies theorem proving (Fikes-Nilsson 1971) form the proof-based viewpoint proof trees are important in carrying out the tasks. A simplified proof tree for AI problem solving is And/Or trees (Nilsson 1969). AND/OR trees Nilsson are trees defined to satisfy a goal from an automated prover stand point.

Formally, a node problem is said to be solved if one of the following conditions hold.

1. The node is the set of terminal nodes (primitive problem—the node has no successor).
2. The node has AND nodes as successors and the successors are solved.
3. The node has OR nodes as successors and any one of the successors is solved.

A solution to the original problem is given by the subgraph of AND/OR graph sufficient to show that the node is solved.

An intelligent AND/OR tree is and AND/OR tree where the tree branches are intelligent trees. It is defined by the author since Nourani (1996d). The branches compute a Boolean

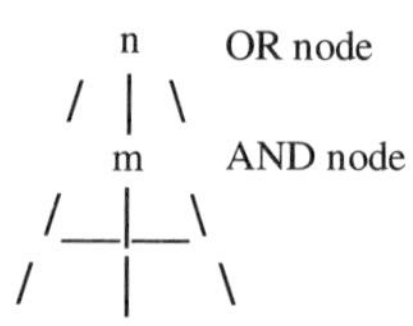

Figure 3. And/or trees, Section 3.

```
         g    OR agent
        / | \
          |
       b1 | b2
          f    AND agent
        /_|_\
       /  |  \
     a1   a2   a3
```

Figure 4. And/or trees, Section 3.

function via agents. Boolean functions are functions on logical predicates to the set $\{0, 1\}$, for true and false, respective. The Boolean function is what might satisfy a goal formula on the tree. An intelligent AND/OR tree is solved iff the corresponding Boolean functions solve the AND/OR trees named by intelligent functions on the trees. Thus node m might be $f(a_1, a_2, a_3)$ & $g(b_1, b_2)$, where f and g are Boolean functions of three and two variables, respectively, and a_i's and b_i's are Boolean valued agents satisfying goal formulas for f and g.

Proof theory is the branch of logic spending its time on the properties of proofs and their trees. First order logic formulas are what can be instantiated at specific structures or AI worlds by facts as to the state of affairs at such structures or worlds. There is a complementary branch of mathematical logic, called model theory, which carries on the logical enterprise from the structures and worlds at which the logics can be instantiated as abstract reality amongst what might "live" at the structures or worlds.

In our projects (Nourani 1995a, 1999c) we also instantiate proof tree leaves with free Skolemized trees. Thus virtual trees are substituted for the leaves. In the present approach, as we shall further define, leaves could be virtual trees.

By a virtual tree we intend a term made of constant symbols and Skolem functions terms. A plan is a sequence of operations in the universe that could result in terms that instantiate the truth of the goal formulas in the universe. That is what goes on as far as the algebra of the model is concerned. It is a new view of planning prompted by our method of planning with GF-diagrams and free Skolemized trees. It is a model-theoretic view. Proof-theoretically a plan is the sequence of proof steps that yields the proof for the goal formula. The proof theoretic view is what the usual AI literature presents.

Computing intelligent multimedia plans with agents at times applies techniques for satisfying several goals simultaneously (Waldinger 1973). There are some starts on planning with multiagent computation (e.g., Muller and Pischel 1994, Nourani 1995c). The author's projects since 1994 have applied multi-board designs to plan with multiagents. The planning process at each stage can make use of GF-diagrams by taking the free interpretation of the possible proof trees that correspond to each goal satisfiability.

The techniques we have applied are to make use of the free Skolemized proof trees in representing plans in terms of generalized Skolem functions. In planning with GF-diagrams that part of the plan which involves free Skolemized trees are carried along with the proof tree for a plan goal. The virtual trees can carry model objects. A function symbol is intelligent iff it is an Agent Functions Set member. To be nontrivial a intelligent function symbol must be defined with a signature that implies message passing between at least two functions in the set, for example, by carrier sharing on the signature. The idea is to do it at abstract syntax trees without grammar specifics.

As an example, suppose I told you I have an academic department with a faculty member which is Superman, and two faculty members which are Swedish speaking, and three which do not talk to anybody not in their expertise area. Without telling you anything else about what they do, I have defined abstract syntax properties. When the signature has few specific agent functions, it implies the signature has defined message paths for them. From the signature a model is defined to assign to abstract syntax trees. The IM hybrid pictures, the message passing actions, and implementing agents are defined by syntactic constructs, with agents appearing as functions. The computation is expressed by an abstract language, for example IMPhora from Chapter 3, that is capable of specifying modules, agents, and their communications.

We have put hybrid picture computing at syntactic constructs that are computed on intelligent trees (Nourani 1993a, 1998a, c, 1999a). The implementing agents, their corresponding objects, and their message passing actions can also be presented by the two-level abstract syntax. For example, IMPhora on agent augmented Visual Java. The agents are represented by function names that appear on the free syntax trees of implementing trees.

The trees defined by the present approach have function names corresponding to computing agents. The computing agent functions have a specified module defining their functionality. A signature defines the language tree compositionality degree and the abstract syntax. Models for the languages are defined by our techniques in Nourani (1995a, 1998c). The Gentzen system defined on MIM (Nourani 1997a) can be assigned an intelligent model theory. The preliminary mathematics is reported in the forthcoming chapters.

EXERCISES

1. Write an AI blocks-world problem solver algorithm with arbitrary objects as blocks applying morphs and trans-morphs, where composition is the "put-on-top" operation with an inverse which is "remove from."
2. Program "intelligence" into your program written for Exercise 1 enclosing with embedded trans-morph routines recognising the frame problem events *a priori* and solving for it.
3. Write an IMPhora program combining Exercises 1 and 2 with two co-operating robots acting on the world.
4. Present a basic intelligent language syntax for two android world as agents co-operating on a blocks-world with n-arbitrary objects. The syntax must specify the android messages necessary to function at the n-object world.
5. Design and implement Exercise 4 with JAVA-to-go.
6. Morph Gentzen computing can be applied to the hybrid pictures to satisfy a plan goal computation. Design and Agent and/or tree for goal satisfaction with hybrid-pictures consisting of frames with three arbitrary objects, taking for granted that hybrid-pictures can be compared for a match with a single operation. The hybrid-picture is a space deck on which there is either a craft P1 or P2, or none; or a window with P1 or P2 visible but not onto the deck. Goal is to put P1 or P2 onto the deck, but not both at the same time.
7. Starting with a visual object with agents presenting a visual filed where there are three object types: spacecraft, planets, and astroids with hybrid pictures and agents as follows:

Hybrid picture 1—Spacecraft A Navigation Window: Agents: A1 Computes available docking times based on the visual field on the window; A2 Carries out docking sequence based on messages to Spacecraft B. Hybrid picture 2 Spacecraft B Navigation Window: Agents: B1 Carries on course based on its visual field window. B2 Accepts and carries out docking manoeuvres from external hovering craft agents. Design a plan to engage docking between A and B at appropriate A and B field windows with agent planning. Morph Gentzen computing can be applied to the hybrid pictures to satisfy a plan goal.

REFERENCES

1. R. Fikes and T. Kehler, *CACM* 9, 904 (1985).
2. R. E. Fikes and N. J. Nilsson, *Artificial Intelligence* 2, 189 (1971).
3. M. Genesereth and Nils Nilsson, "Logical Foundations of Artificial Intelligence." Morgan-Kaufmann, 1987.
4. G. Gentzen, "Beweisbarkeit und Unbewiesbarket von Anfangsfallen der trasnfininten Induktion in der reinen Zahlentheorie." *Math. Ann.* 119, 140 (1943).
5. K. J. Holyoak and P. Thagard, "Mental Leap, Analogy in Creative Thought." MIT Press, Cambridge.
6. S. Kleene, "Mathematical Logic." Introduction to Metamathematics, John Wiley and Sons, Van Nostrand, NJ, 1952.
7. M. Minsky, "The Society of Mind." Simon-Schuster, 1986.

8. J. P. Muller and M. Pischel, "Modeling Interactive Agents in Dynamic Environments," Proceedings 11th European Conference on AI. John Wiley and Sons, Amsterdam, The Netherlands, 1994.
9. N. J. Nilsson, "Searching, Problem Solving, and Game-Playing Trees for Minimal Cost Solutions" (A. J. Morell, Ed.), Vol. 2, pp. 1556–1562. IFIP, North-Holland, Amsterdam, 1969.
10. N. J. Nilsson, "Principles of Artificial Intelligence." Morgan Kaufmann, 1980.
11. C. F. Nourani, "Planning and Plausible Reasoning in AI," pp. 150–157. Proc. Scandinavian Conference in AI, IOS Press, Roskilde, Denmark, 1991.
12. C. F. Nourani, "Slalom Tree Computing," AI Communications, The European AI Journal. IOS Press, Amsterdam, 1993a.
13. C. F. Nourani, "Free Proof Trees and Model-Theoretic Planning." AISB, Sheffield, 1995a.
14. C. F. Nourani, "Double Vision Computing," Intelligent Autonomous Systems. Karlsruhe, Germany, 1995b.
15. C. F. Nourani, "Multiagent Robot Supervision," Proceedings ECML Learning Robots Track. Heraklion, 1995c.
16. C. F. Nourani, "Intelligent Multimedia—A New Computing Technique and New AI Paradigms," International Conference on Computational Intelligence and Multimedia Applications. Victoria, Australia, 1996a.
17. C. F. Nourani, "Linguistics Abstraction," Proceedings ICML96—International Conference on Mathematical Linguistics. Catalunya, Tarragona, Spain, 1996b.
18. C. F. Nourani, "A Multiagent Cognitive Theory," International Conference Cognitive Science. Lawrence Earlbaum Associate Publishers, Mahawk, New Jersey, 1996c.
19. C. F. Nourani, "Multiagent Chess Games," AAAI Chess Track, Providence, RI, 1996d.
20. C. F. Nourani, "MIMLogik," Summer Logic Colloquium, BSL, Quarterly Publication by the Association for Symbolic Logic, Inc., Prague, 1997a.
21. C. F. Nourani, "AII and Heterogeneous Software Design," Eighth European Workshop on Modeling Autonomous Agents in a Multi-Agent World. University of Karlskrona/Ronneby, Ronneby, Sweden, 1997b.
22. C. F. Nourani, "Creating Art and Motion Pictures with Intelligent Multimedia," Written for the Artbyte Magazine, 1998.
23. C. F. Nourani, "Intelligent Trees, Thought Models, and Intelligent Discovery," Model-based Reasoning in Scientific Discovery. Pavia, Italy, 1998a.
24. C. F. Nourani, "Intelligent Trees and Consciousness Science," Proceedings International Conscience in Science and Philosophy Conference. Charelston, Illinois, 1998b.
25. C. F. Nourani, "Intelligent Languages—A Preliminary Syntactic Theory" (Jozef Gruska and Jiri Zlatuskasp, Eds.), Vol. 1450. Mathematical Foundations CS, Grammatics, Lecture Notes in Computer Science, Springer, 1998c.
26. C. F. Nourani, "Intelligent Multimedia—New Computing Techniques and Applications" (Ch. Freytag and V. Wolfengagen, Eds.), Proceedings of 1st International Workshop on Computer Science and Information Technologies. MEPhI Publishing, Moscow, Russia, 1999a.
27. C. F. Nourani, "Autonomous Double Vision Spacecraft and MultiAgent Supervision," Proceedings Conference, Agent Control Autonomy Track. Seattle, WA, 1999b.
28. C. F. Nourani, "Morph Gentzen Virtual Computing, East Coast Algebra Day Conference." North Carolina, 1999c.
29. C. F. Nourani, "Virtual World Models, Illusion, and Morphs," Sank Augustin, Germany, IMK, 2000.
30. R. Waldinger, "Achieving Several Goals Simultaneously," SRI Technical Note 107, Project 2245. Menlo Park, CA.
31. P. H. Winston, "The Psychology of Computer Vision." McGraw Hill, New York, 1975.

CHAPTER 5

Agent Computing and Intelligent Multimedia

CONTENTS

ISBN: 1-58883-037-3/$35.00

Intelligent Multimedia Computing Science
By Cyrus F. Nourani
Pages: 47–71

1. INTRODUCTION

This chapter introduces agent computing with interactive intelligent multimedia. An overview to a practical agent computing model based on beliefs, intentions, and desire is presented and possible augmentation to intelligent multimedia is explored. Nilsson-Genesereth (1987) introduces agent architectures. There is an inevitable mathematical interplay here as a computing science which is further presented on specific chapters on the mathematical foundations to intelligent multimedia and the mathematical basis to agent computing. Thus the reader benefits from a gentle presentation while glancing the mathematical chapters when necessary.

The term "agent" is applied to refer to AI constructs that enable computation on behalf of an AI activity. It also refers to computations that take place in an autonomous and continuous fashion, while considered a high-level activity, in the sense that its definition is software/hardware, thus implementation, independent. Such functionality is typical of what is required to implement complex autonomous planning systems. Agents are in most cases informable, thus allowing message passing actions. Thus cooperating expert systems can be viewed as simple examples for the agent models of AI systems. Concurrent world populated by agents having many concurrent goals present new planning problems. Neither a purely reactive nor standard planning might apply. The former does not apply because the agents have to know what they will be doing in the future to make decisions what to do in the present. The latter does not fully apply because of the size and the scope of the problems presented. Multilevel multiagent hybrid planning has to be defined. The project applies knowledge representation on generalized diagrams. The method of generalized diagrams, abbreviated by G-diagram, from Nourani (1993, 1988, 1995) were invented for AI planning and reasoning, formulating various notions of generalized and free diagrams. A specific agent might have internal state set I, which the agent can distinguish its membership. The agent can transit from each internal state to another in a single step. With our multi-board model agent actions are based on I and board observations. There is an external state set S, modulated to a set T of distinguishable subsets from the observation viewpoint.

A sensory function $s : S \rightarrow T$ maps each state to the partition it belongs. Let A be a set of actions which can be performed by agents. A function action can be defined to characterise an agent activity action: $T \rightarrow A$. There is also a memory update function mem: $I \times T \rightarrow I$.

Dynamics and situation compatibility is introduced as a structural way to compute and compare epistemic states. Worlds, epistemics, and cognition for androids are introduced with precise statements. The foundations are applied to present a brief on Computational Illusion, affective computing, and Virtual Reality. KR for AI Worlds, and Computable Worlds are presented with diagrams. Cognitive modeling is briefed with an introduction to ordinal dynamics. A preview to computational epistemology with cardinality and concept descriptions are introduced. Deduction models and perceptual computing are presented with a new perspective. Intelligent multimedia interfaces are an important component to the practical computational aspects. Visual context and objects are presented with multiagent intelligent multimedia. Context abstraction and meta-contextual reasoning are introduced as a new field. Multiagent visual multi-board planning is introduced as a basis to intelligent multimedia with applications to spatial computing.

2. THE AGENT MODELS AND DESIRE

Let us start with the popular agent computing model the Beliefs, Desire, and Intentions, henceforth abbreviated as the BID model (Brazier-Truer et al.). BID is a generic agent computing model specified within the declarative compositional modeling framework for multi-agent systems, DESIRE. The model, a refinement of a generic agent model, explicitly specifies motivational attitudes and the static and dynamic relations between motivational attitudes. Desires, goals, intentions, commitments, plans, and their relations are modeled.

Different notions of strong and weak agency are presented at Wooldridge and Jennings (1995). Velde and Perram (1996) distinguished big and small agents. To apply agent computing with intelligent multimedia some specific roles and models have to be presented for agents. The BID model has emerged for a "rational agent": a rational agent described using cognitive notions such as beliefs, desires and intentions. Beliefs, intentions, and commitments play a crucial role in determining how rational agents will act. Beliefs, capabilities, choices, and commitments are the parameters making component agents specific. Such bases are applied to model and to specify mental attitudes (Shoham 1993, Rao and Georgeff 1991, Cohen and Levesque 1990, Shoham 1991, Dunin-Keplicz and Verbrugge 1996).

A generic BID agent model in the multiagent framework DESIRE is presented towards a specific agent model. The main emphasis is on static and dynamic relations between mental attitudes, which are of importance for cooperative agents. DESIRE is the framework for design, and the specification of interacting reasoning components is a framework for modeling, specifying and implementing multi-agent systems, see Brazier et al. (1995, 1996), Dunin-Keplicz and Treur (1995). Within the framework, complex processes are designed as compositional architectures consisting of interacting task-based hierarchically structured components.

The interaction between components, and between components and the external world is explicitly specified. Components can be primitive reasoning components using a knowledge base, but may also be subsystems which are capable of performing tasks using methods as diverse as decision theory, neural networks, and genetic algorithms. As the framework inherently supports interaction between components, multi-agent systems are naturally specified in DESIRE by modeling agents as components.

The specification is sufficient to generate an implementation. Specific techniques for such claims might be further supported at Nourani (1993a, 1999a). A generic classification of mental attitudes is presented and a more precise characterization of a few selected motivational attitudes is given. The specification framework DESIRE for multi-agent systems is characterized. A general agent model is described. The framework of modeling motivational attitudes in DESIRE are discussed.

2.1. Mental Attitudes

Agents are assumed to have the four properties required for the weak notion of agency described in Wooldridge and Jennings (1995). Thus, agents must maintain interaction with their environment, for example observing and performing actions in the world: reactivity; be able to take the initiative: pro-activeness; be able to perform social actions like communication, social ability; operate without the direct intervention of other (possibly human) agents: autonomy.

Four main categories of mental attitudes are studied in the AI literature: informational, motivational, social and emotional attitudes. The focus is on motivational attitudes, although other aspects are marginally considered. In Shoham and Cousins (1994), motivational attitudes are partitioned into the following categories: goal, want, desire, preference, wish, choice, intention, commitment, plan. Individual agents are assumed to have intentions and commitments both with respect to goals and with respect to plans.

A generic classification of an agent's attitudes is defined as follows:

1. Informational attitudes: Knowledge; Beliefs.
2. Motivational attitudes: Desires; Intentions—Intended goals and Intended plans.
3. Commitments: Committed goals and Committed plans.

In planning, see Section 6, the weakest motivational attitude might be desire: reflecting yearning, wish and want. An agent may harbor desires that are impossible to achieve. Desires may be ordered according to preferences and, as modeled in this paper, they are the only motivational attitudes subject to inconsistency. At some point an agent must just settle on a limited number of intended goals, i.e., chosen desires.

2.2. Specifying Beliefs, Intentions, Desire Model Agents

The BID-architectures upon which specifications for compositional multi-agent systems are based on the result of analysis of the tasks performed by individual agents and group of agents. Task (de)compositions include specifications of interaction between subtasks at each level within a task (de)composition, making it possible to explicitly model tasks, which entail interaction between agents. The formal compositional framework for modeling multi-agent tasks DESIRE is introduced here. The following aspects are modeled and specified:

(1) a task (de)composition, (2) information exchange, (3) sequencing of (sub)tasks, (4) subtask delegation, (5) knowledge structures.

Information required/produced by a (sub)task is defined by input and output signatures of a component. The signatures used to name the information are defined in a predicate logic with a hierarchically ordered sort structure (order-sorted predicate logic). Units of information are represented by the ground atoms defined in the signature. The role information plays within reasoning is indicated by the level of an atom within a signature: different (meta) levels may be distinguished. In a two-level situation the lowest level is termed object-level information, and the second level meta-level information.

Some specifics and a mathematical basis to such models with agent signatures might be obtained from Nourani (1996a) where the notion had been introduced since 1994 and Chapter 6. Meta-level information contains information about object-level information and reasoning processes; for example, for which atoms the values are still unknown (epistemic information). Similarly, tasks that include reasoning about other tasks are modeled as meta-level tasks with respect to object-level tasks. Often more than two levels of information and reasoning occur, resulting in meta-meta-information and reasoning.

Information exchange between tasks is specified as information links between components. Each information link relates output of one component to input of another, by specifying which truth-value of a specific output atom is linked with which truth value of a specific input atom. For a multiagent object information exchange model see, for example, Nourani (1996a). The generic model and specifications of an agent described above can be refined to a generic model of a rational BID-agent capable of explicit reasoning about its beliefs, desires, goals and commitments.

3. DYNAMICS AND SITUATIONS

3.1. Worlds and A Robot's Touch

Starting with the issues raised by Hiedegger in 1935–1936, starting with the notion of "What is a thing" as put forth in Heidegger (1963). The author's immediate reaction when presented with such challenges to computing applications with philosophical epistemics, while visiting INRIA, Paris around 1992, was to start with "first principles," not touching such difficult areas of philosophy and phenomenology, and only present views to what they could imply for the metamathematics of AI. However, since the author's techniques were intended for AI computations and reasoning, rather than knowledge representation from observations, as it is the case in Didday (1990), Heidegger's definitions had to be taken further. The common point of interest is symbolic knowledge representation. However, the

```
Person ------Cognition----Real Worlds and Objects
|                                               |
|                                               |
|                                               |
|--------------\ \------------------------------|
|         The Sensory Illusion Gap              |
|                                               |
|                                               |
Robot------World Description---RealWorlds Objects
```

Figure 1. The sensory illusion gap.

research directions are two essentially orthogonal, but not contradicting, views to knowledge representation.

3.2. Computational Illusion and Virtual Reality

der Vielleicht Vorhandenen objects are the perhaps computable and might be a computational illusion, as further illustrated by the following figure on the human intelligence and artificial intelligence comparison. Thus the robot's senses are not always real. The important problem is to be able to define worlds minimally to have computable representations with mathematical logic thus the ability to make definitive statements. Heidegger's Die Frage nach dem Ding will prove to be a blessing in disguise. Could it have computing applications to things without.

Heidegger had defined three sorts of things

1. Things in the sense of being "within reach," des Vorhandenen.
2. Things which "unify" things of the first kind, or are reflections on, resolution and actions.
3. Things of kind 1 or 2 and also any kind of things which are not nothing.

To define a logic applicable to planning for robots reaching for objects, the der Vielliecht Vorhandenen computational linguistics game is defined. To start, let us explore Heidegger's views of the "des Vorhandenen," having to do with what object is within "reach" in a real sense. In AI and computing applications notion of des Vorhandnen is not absolute.

As an AI world develops the objects that have names in the world are at times des Vorhandnen and as defined by a principle of Parsimony only des Vorhandnen in an infinitary sense of logic Nourani (1984, 1991). The logical representation for reaching the object might be infinitary only. The phenomenological problem from the robot's standpoint is to acquire a decidable descriptive computation for the problem domain.

Thus what is intended to be reached can stay always out of reach in a practical sense, unless it is at least what I call der Vielliecht Vorhandenen Nourani (1994a, b). The computing issue is the artificial intelligence computation and representation of real objects. That is, we can make use of symbolic computation to be able to "get at" a real object. At times, however, only infinite computations could define real world objects. For example, there is a symbolic computation for an infinite ordinal, by an infinite sequence of successor operations on 0.

Furthermore, the present notion of der Vielliecht Vorhandenen is not intend to be the sense in which a robot cannot reach a particular object. The intent is that the language could have names for which the corresponding thing is not obvious in the AI world and there is incomplete information until at some point the world is defined enough that there is a thing corresponding to a name, or that at least there is a thing by comprehension, which only then becomes des Vorhandnen as the AI world is further defined or rearranged. These issues are examined in the computational context in the sections below.

For example, the der Vielleicht Vorhandenen game has a winning strategy if the world descriptions by G-diagrams define the world enough to have a computation sequence to reach for an intended object. This implies there must be a decidable descriptive computation (Nourani 1994, 1996) for the world applied. The immediate linguistics example of these concepts from natural languages is a German child's language in which to "vor" and "handenen" are some corresponding things in the child's language world and mind, but "vorhandenen" is not a thing in that child's world and only becomes a thing as the linguistics world is further defined for the child. When can the child reach for the stars? As Heidegger implies "Fur das kind in Menschen bleibt die nacht die Naherin der Sterne."

The same problem might arise when the robot tries to actually get at elementary objects, where the robot finds what is called a paradox in Didday (1990): that elementary objects have to be defined by comprehension. Comprehension is a closure with respect to properties that are essential and cannot be dropped without loss to the enclosed. Since the paper in its theory that is presented in part here, does not restrict Heidegger's definition, it can be further developed for AI applications. I might suggest ways of incorporating the above for computing applications. The problems with words, objects and symbols have been there since Quine (1950's).

3.3. Representing AI Worlds

Diagrams are the "basic facts of a model," i.e. the set of atomic and negated atomic sentences that are true in a model. Generalized diagrams are diagrams definable by a minimal set of functions such that everything else in the model's closure can be inferred, by a minimal set of terms defining the model. Thus providing a minimal characterisation of models, and a minimal set of atomic sentences on which all other atomic sentences depend.

However, since we cannot represent all aspects of a real world problem, we need to restrict the representation to only the relevant aspects of the real world we are interested in. Let us call this subset of relevant real world aspects *the AI world.* Our primary focus will be on the relations amongst KR, AI worlds, and the computability of models. Truth is a notion that can have dynamic properties. Interpretation functions map language constructs (constants, function and predicate symbols) onto entities of the world, and determine the notion of truth for individuals, functions and relations in the domain.

The real world is infinite as the AI worlds are sometimes. We have to be able to represent these ideas within computable formulations. Even finite AI worlds can take an exponential number of possible truth assignments. Thus the questions: how to keep the models and the KR problem tractable, such that the models could be computable and within our reach, are an important area (Nourani 1991, 1993a, b, 1994, 1996, Lake 1996).

3.4. Computable World Models

To prove Godel's completeness theorem, Henkin defined a model directly from the syntax of the given theory. The reasoning enterprise requires more general techniques of model construction and extension, since it has to accommodate dynamically changing world descriptions and theories. The techniques in Nourani (1983, 1987, 1991) for model building as applied to the problem of AI reasoning allows us to build and extend models through diagrams. We apply generalized diagrams to define models with a minimal family of generalized Skolem functions. The minimal set of function symbols are those with which a model can be built inductively. The models are computable as proved by Nourani (1984, 1993a, 1995b). The G-diagram methods applied and further developed here allows us to formulate AI world descriptions, theories, and models in a minimal computable manner. Thus models and proofs for AI problems can be characterized by models computable by a set of functions.

3.5. AI Model Diagrams

An AI world consists of individuals, functions on them, and relations between them. These entities allow us to fix the semantics of a language for representing theories about AI worlds. We take the usual model-theoretical way, and assign via an interpretation function individuals to constants, functions to function symbols and relations to predicate symbols. Let us define a simple language $\mathscr{L} = \langle$\{tweety\},\{a\},\{bird\}, predicate letters, and FOL$\rangle$. A model may consist of \{bird(tweety), ¬ penguin(tweety) → bird(tweety), bird(tweety) v ¬ bird(tweety), ... \}, others may consist of $\{p(a), \neg p(a) \rightarrow p(a), p(a)vp(x), p(a)vp(x)vp(y), \ldots\}$.

Because we can apply arbitrary interpretation functions for mapping language constructs into AI worlds, the number of models for a language is infinite. Although this makes perfect sense from a theoretical and logical point of view, from a practical point of view, this notion of model is too general for AI applications.

For AI we want effective and computable models. Thus, it is useful to restrict the types of models that we define for real world applications. Primarily, we are interested in models with computable properties definable from a theory. In order to point out the use of the generalized method of diagrams we present a brief view of the problem of planning form (Nourani 1991) within the present formulation. The diagram of a structure in the standard model-theoretic sense is the set of atomic and negated atomic sentences that are true in that structure (see Chapter 6). The generalized diagram (G-diagram) is a diagram in which the elements of the structurc arc all represented by a minimal family of function symbols and constants.

It is sufficient to define the truth of formulas only for the terms generated by the minimal family of functions and constant symbols. Such assignment implicitly defines the diagram. This allows us to define a canonical model of a theory in terms of a minimal family function symbols. Models uphold to a deductive closure of the axioms modelled and some rules of inference, depending on the theory. By the definition of a diagram they are a set of atomic and negated atomic sentences. Hence a diagram can be considered as a basis for a defining model, provided we can by algebraic extension, define the truth value of arbitrary formulas instantiated with arbitrary terms.

Thus all compound sentences build out of atomic sentences then could be assigned a truth-value, handing over a model. This will be made clearer in the following subsections. The following examples would run throughout the paper. Consider the primitive first order language (FOL)

$$\mathcal{L} = \{c\}, \{f(X)\}, \{p(X), q(X)\}$$

Let us apply Prolog notation convention for constants and variables and the simple theory {for all X: $p(X) \rightarrow q(X)$, $p(c)$}, and indicate what is meant by the various notions.

$$\begin{aligned}(\text{model}) = {} & \{p(c), q(c), q(f(c)), q(f(f(c))), \ldots\}, \\ & \{p(c)\&q(c), \ldots p(c)\&p(X), p(c)\&p(f(X)), \ldots\}, \\ & \{p(c)vp(X), p(c)vp(f(X)), p(c) \rightarrow p(c)\ldots\}. \\ (\text{diagram}) = {} & \{p(c), q(c), p(c), q(f(c)), q(f(f(c))), \ldots\}, \ldots, q(X)\};\end{aligned}$$

i.e., the diagram is the set of atomic formulas of a model. Thus the diagram is

$$(\text{diagram}) = \{p(c), q(c), q(f(c)), q(f(f(c))), \ldots, q(X)\}.$$

There are various notions of diagram from the author's papers (see references) applied here.The term generalized diagram refers to diagrams that are instantiated with generalized Skolem functions. The generalized Skolem functions were defined in by the author, for example, Nourani (1991) as functions with which initial models are defined inductively. We can define generalized diagrams based on the above. The term generalized is applied to indicate that such diagrams are defined by algebraic extension from basic terms and constants of a language. The diagrams are completely defined from only a minimal function set.

Generalized diagrams is

$$(\text{generalized diagram}) = \{p(c), q(c), p(f(t)), q(f(t))\}$$

for t defined by induction, as $\{t_0 = c$, and $t_n = \{f(t(n-1))\}$ for $n > 0$. It is thus not necessary to redefine all $f(X)$'s since they are instantiated.

3.6. Cognitive Modelling

Cognitive modelling can be enhanced with diagrams since our G-diagram techniques imply automatic models from basic functions. A systematic methodology for cognitive modelling can be considerably assisted by the G-diagram modelling. The area has been emphasised by Cooper et al. (1996). The notion of a symbolic object is put forth in Didday (1990) by considering some individuals as elementary things and then defining symbolic objects from the elementary objects by a comprehension technique with some descriptor functions. Thus comprehension and descriptor make the jump from elementary objects to symbolic objects functions.

So far as the issues with symbolic objects are concerned there is a correspondence to the approach with Generalized Diagrams that can be defined. In our earlier papers the method of Possible Worlds is captured by that of the definition of generalized nondeterministic diagrams. Further, the earlier notion of a set {T, F, X} in Nourani (1991) and diagrams with generalized Skolemization in a recent paper of this author Nourani (1993a, 1995b) handle arbitrary valued logic. Such correspondence could be subject of forth coming papers.

There are various issues having to do with correspondence of symbolic objects and real world things to address. If we were to search for a model-theoretic (Nourani 1991) view of these in view of a triple $\langle \mathcal{L}, |\mathcal{A}|, \mathcal{A} \rangle$ for a proper language $\mathcal{L}$, we could gain some insight to the approaches. The other components of the triple are the concept of a model $\mathcal{A}$ and its universe $|\mathcal{A}|$, see for example Kleene (1967).

The diagram of a structure is the set of atomic and negated atomic sentences that are true in that structure. The generalized diagram (G-diagram) (Nourani 1987, 1991) is a diagram in which the elements of the structure are all represented by a minimal family of function symbols and constants, such that it is sufficient to define the truth of formulae only for the terms generated by the minimal family of functions and constant symbols.

Such assignment implicitly defines the diagram. It allows us to define a canonical model of a theory in terms of a minimal family of function symbols. Generalized diagrams are precisely what allow us to build models from the syntax of a theory, thus allow for symbolic computation of models and theories. The author had defined the notion of generalized diagram eversince (Nourani 1984) for AI reasoning.

Since the author has shown generalized diagrams for models capture the possible worlds formulation in a concise and elegant manner. In a possible world approach one focuses on the "state of affairs" that are compatible with what one knows to be true. We have shown in the above papers how the approach with G-diagrams to possible worlds gives an implicit treatment to modalities.

The correspondence of modalities to Possible Worlds and the containment of the possible worlds approach by the generalized diagrams approach of this author implies that we can present a model-theoretic formulation of the concept of modal symbolic objects (Didday 1990, Nourani 1992, 1993b). Objects with varying properties, with cross product of modes formed from various generalized diagrams corresponding to each mode. Also the notion of language $\mathcal{L}$ has some consequences as far as the model theory to be developed is concerned. Then all the notions for the various modes could be defined and perhaps open new views of computation on generalized diagrams allowing us to represent views of cognition and computation with modes of thought in artificial intelligence.

3.7. Extensions and Models

In Nourani (1984, 1991) we have shown how to characterise AI computations by model extensions that are defined by theories with nonmonotonic (AI 1980) dynamics. This direction of research could apply to symbolic knowledge representation as well. Didday defines the I-extension to symbolic objects by defining extensions to a mode. It then becomes possible to extend definitions, properties, and qualities of modal symbolic objects. The applicability of the formulations in this area could be further pursued. A point of observation is the definition of completeness: a symbolic object is said to be complete if and only if the properties that characterize its extension are exactly those, whose conjunction defines the object.

Nourani (1991) has applied possibility theory with plausible beliefs are closed under finite conjunction, and that probabilistic belief does not have this finite conjunction property. We also showed that our approach has the infinite conjunction property, i.e., beliefs are closed under infinite conjunction. We expect our mathematical approach to reasoning to have some relevance to defining I-extensions and their completeness properties. But this brief statement could take up a wonderful research project to live up to its expectations. The relations of our papers to the present notions are not within the scope of the present paper. Some preliminary concepts are put forth in the following sections to the direction of research.

3.8. Situations and Possible Worlds

What the dynamic epistemic computing defines is not a situation logic in the exact Barwise sense (Barwsie 1985a, b). The situation and possible worlds concepts are the same. However, we define epistemics and computing on diagrams, with an explicit treatment for modalities.

The treatment for modalities are similar to Hintikka's Model Sets (Hintikka 1963, Nourani 1991). A possible world may be thought of as a set of circumstances that might be true in an actual world.

The possible worlds analysis of knowledge began with the work of Hintikka through the notion of model set and (Kripke 1963) through modal logic, in which rather than considering individual propositions, one focuses on the "state of affairs" that are compatible with what one knows to be true, rather than being regarded as possible, relative to a world believed to be true, rather than being absolute. For example, a world w might be a possible alternative relative to w', but not to w''.

Possible worlds consists of a certain completeness property: for any proposition p and world w, either p is true in w or not p is true in w. Note that this is exactly the information contained in a generalized diagram, as defined in the previous section. Let W be the set of all worlds and p be a proposition. Let (p) be the set of worlds in which p is true. We call (p) the truth-set of P. Propositions with the same truth-set are considered identical. Thus there is a one-one correspondence between propositions and their truth sets. Boolean operations on propositions correspond to set-theoretic operations on sets of worlds. A proposition is true in a world if and only if the particular world is a member of that proposition.

3.9. Epistemic States and Ordinal Dynamics

Generalized diagrams, possible worlds and their logic, model theory, and set theory, are applied to put forth a basis for quantifying epistemology and put forth a computability theory for epistemology. There is, however a theoretical development in philosophical logic towards quantifying the dynamics of epistemics. This author has defined an artificial intelligence planning theory, with applications to robot planning, which applies epistemic with automated deduction. Logics for cognition epistemics in a robot computing theory logically corresponds to designing computing techniques to implement the computational epistemic's dynamics.

The robot computing theory we have defined in Nourani (1991) is an example of a computational epistemics theory for robot planning and it runs on generic ordinal coded diagrams for models. Gardenfors (1988) applies epistemics to knowledge in flux as logic for cognition. Suppose we represent a robot's beliefs by a set of propositions, which are a set of subsets of W. If the robot already believes a proposition p, then the new information does not affect its belief state. If it does not believe p, then the impact of new information on the robot's belief state has to be defined. Sphon epistemics carrys on the revision problem through a partitioning of possible worlds. Each possible world is assigned an ordinal representing its degree of implausibility.

The higher the assigned ordinal the more implausible that world as an actual world. Let k be the function assigning ordinals to each world. Sphon calls this function an ordinal conditional function (OCF). The set of worlds w for which $k(w) = 0$ is the set of most plausible worlds. The robot believes that the real world is a member of $k_{i(0)}$, where k_i is the inverse of k, and considers no world as plausible as any world in that set. The robot believes proposition p iff $k_{i1(0)}$ is a subset of p, i.e. iff p is true in all the most plausible worlds. The function k can be extended to propositions as follows, where p is a proposition, we let

$$k(p) = \min\{k(w) | w \text{ is in } p\}.$$

The strength of belief in a proposition p is represented by the degree of implausibility of $\neg p$. The more implausible $-p$ is, stronger the belief in p. Hence we may define Bf, the belief function, by Bf$(p) = k(\neg p)$. If $k(p)$ and $k(\neg p)$ happened to be zero, then there is no degree of belief in either. In the present theory of belief revision, when the robot comes to believe a proposition p that it does not already believe, the result is a new ranking of the possible worlds. If a is the strength with which the new proposition p is believed, then the new ordinal conditional function, representing the robot's new belief state, is denoted by $k(p, a)$. $k(p, a)$ is defined as follows:

$$\begin{aligned} k(p, a)(w) &= k(w) - k(p), \quad \text{if } w \text{ is in } p \\ &= a + k(w) - k(\neg p), \quad \text{if } w \text{ is in } \neg p \end{aligned}$$

The papers show how to solve well-known planning problems with the above. Further bases and applications are reported at Williams (1994). A fundamental problem in KR is inherent intractability to comply with KB with limited belief and knowledge. The G-diagram techniques are an alternate way to formally specify beliefs with deductive and limited yet fully introspective KB. The area is since been viewed by Lakemeyer (1996). The computability problems in KR are further treated by Nourani (1996b).

3.10. Generic Encoding and Epistemology

From a formal representation of epistemic states as presented by Sphon (1988), the generalized diagram formulation of possible worlds, and the encoding of epistemic states by Nourani (1987, 1991) we have the following conclusions. That Probabilistic Epistemology (PE) corresponds to intuitive notions of subjective and objective probability.

It appears that Deterministic Epistemology (DE) leads to truth values for propositions, and belief by some epistemic subject that a proposition is true, false, or neither: thus to the notion of "truth." In Nourani (1991) we have shown that the notion of truth or belief in deterministic epistemology is closed under infinite conjunction, whereas this is not true of probabilistic epistemology.

Let Deterministic Epistemology to be the logic and epistemics definable by a deterministic logic and model theory, i.e., the known standard logic and model theory, allowing for infinitary logics. Let us further define Probabilistic Epistemology (PE), to be a logic and epistemics defined by probabilities, as in known probabilistic logics. Probabilistic Epistemology corresponds to intuitive notions of subjective and objective probability. Remarks—It might seem that the ranking of the worlds with ordinals and the OCF approach corresponds to the rankings of the worlds in terms of their probability, with the most probable world having rank 0, the next most probable world rank one, and so on. However, such an easy correspondence runs into difficulties. In Sphon's formulation, a proposition is believed just in case it is entailed by (i.e. a superset of) the set of worlds of rank zero. This implies that belief is closed under conjunction. For if the set of most plausible worlds entails each member of a finite set of propositions, then it also entails their conjunction. Having defined DE and PE the following theorem is stated as an example on the area. The mathematical specifics are treated in the foundational chapters.

THEOREM 5.1. *There is no reduction from DE to PE.*

What the theorem means is that epistemic computations defined by DE are not always reducible to those for PE in the computability theory sense. It is not a question of polynomial reducibility—it is reducibility at all. This is not obvious at all if you think about it. It implies we have a stronger computability degree with deterministic epistemology, infinitary logic, model theory and set theory for computational epistemology. Borrowing the following definition from Chapter 6 let us start with some specifics on diagrammatic computing.

A generic diagram or *G-diagram* for a structure M is a diagram where there is a specified function set with which all the structure's language formulas truth values can be encoded with respect to the structure M. The specified functions above are those by which a standard model could be defined. Examples for such specified functions appear at set theory and foundations, e.g., $\Sigma 1$ Skolem functions. The foundations chapters prove that G-diagrams for models can encode possible worlds.

Now let us examine the definition of situation and view it in the present formulation. A situation consists of a nonempty set D, the domain of the situation, and two mappings: g, h. g is a mapping of function letters into functions over the domain as in standard model theory. h maps each predicate letter, p_n, to a function from D_n to a subset of $\{t, f\}$, to determine the truth value of atomic formulas as defined below The logic has four truth values: the set of subsets of $\{t, f\}.\{\{t\}, \{f\}, \{t, f\}, 0\}$, the latter two corresponding to inconsistency, and lack of knowledge of whether it is true or false.

Due to the above truth values the number of situations exceeds the number of possible worlds. The possible worlds being those situations with no missing information and no contradictions. From the above definitions the mapping of terms and predicate models extend as in standard model theory. Next, a compatible set of situations is a set of situations with the same domain and the same mapping of function letters to functions. In other worlds, the situations in a compatible set of situations differ only on the truth conditions they assign to predicate letters.

The dynamics of epistemic states as formulated by generalized diagrams is exactly what addresses the compatibility of situations. How an algebra and model theory for epistemic states is to be defined by generalized diagram of possible worlds, is exactly what Nourani (1987, 1991) leads us to. To decide compatibility of two situations we compare their generalized diagrams. Thus the foundations chapters prove that: Two situations are compatible iff their corresponding generalized diagrams are compatible with respect to the Boolean structure of the set to which formulas are mapped (by the function *h* above, defining situations).

One of the implications of the above towards cognition and descriptive computing from the point of view of computer vision is the notion of der Vielliecht Vorhandene. It is within the mathematical expressive power of our methods (Nourani 1991, 1994) with infinitary logic to form an infinite conjunction of beliefs with respect to an AI world. Thus we can represent an AI world and all the compatible generalized diagrams that can make "something" des Vorhandene form the model-theoretic point of view and descriptive computing. It further allows us to compare dynamics *a priori* based on personality descriptions and a specific movie script to precast critical movie and TV scenarios with interactive intelligent multimedia, and emotional agent computing.

But the cognition dimension only relies on observable data and cannot form a conjunction of beliefs on every sample of data to conclude that the same "something" above is des Vorhandenen. This is a consequence of the above theorems and formulation. The analogy is that of proof theory, model theory and Godel's incompleteness theorem Kleene (1967).

3.11. Cardinality and Concept Descriptions

Let us present what we refer to as Descriptive Computation applying generalized diagrams, following our earlier papers Nourani (1988, 1991). We define descriptive computation to be computing with G-diagrams for the model and techniques for defining models with G-diagrams from the syntax of a logical language. G-diagrams are diagrams definable with a known function set. Thus the computing model is definable by G-diagrams with a function set.

The analogous terminology in set theory refers to sets or topological structure definable in a simple way. Thus by descriptive computation we can address artificial intelligence planning and theorem proving, for example. The author in Nourani (1984) pursues the latter computational issues. The logical representation for reaching the object might be infinitary only. We show in Nourani (1994a, b, 1996) that the artificial intelligence problem from the robot's stand point is to acquire a decidable descriptive computation for the problem domain. Nourani (1996) proves specific theorems for descriptive computing on diagrams. A compatibility theorem applies descriptive computing to characterise situation compatibility. Further, a computational epistemic reducibility theorem is proved by the descriptive computing techniques on infinitary languages by the author in Nourani (1994b). A deterministic epistemics is defined and it is proved not reducible to known epistemics. Cardinality restrictions on concepts are important areas explored by AI. The concept description logics systems allow users to express local cardinality on particular role filers. Global restrictions on the instances of a concept are difficult and not possible. Cardinality restrictions on concepts can be applied as an application domain description logic (Baader et al. 1996). The concept definitions with G-diagrams for localized KR and its relations to descriptive computable sets can be applied to concept cardinality restriction. By applying localized functions to define G-diagrams models for languages as defined by Baader et al. (1996) can be generated with cardinality restrictions.

3.12. Deduction Models and Perceptual Computing

It might be illuminating to compare the G-diagram techniques and computational epistemology to the (Konolige 1984) starting with the consequential closure problem for artificial intelligence and the possible worlds. What Konologie starts with the infeasibility premise for consequential closure, i.e. the assumption that an agent knows all logical consequences of his beliefs. The deductive model is defined for situations where belief derivation is logically incomplete. The area had been voiced since Fodor (1975) and Moore (1980).

Konolige applies a model where beliefs are expressions in the agent's "mind" and the agent reasons about them by manipulating syntactic objects. When the process of belief derivation is logically incomplete, the deduction model does not have the property of the consequential closure. Konolige defines a saturated deduction model and claims a correspondence property: For every modal logic of belief based on Kripke possible world models, there exists a corresponding deduction model logic family with an equivalent saturated logic.

In Nourani (1984, 1987, 1991, 1995, 1996) and the present chapter it is shown that there is a minimal characterization of AI reasoning models with generic diagrams from which models can be defined for belief revision and automatically generated. The G-diagrams are defined for incomplete KR, modalities, and model set correspondence. What computational epistemology defines is a model theoretic technique whereby without the consequential closure property requirements on agents a model-theoretic completeness can be ascertained via non-deterministic diagrams. Specific modal diagrams were defined for computational linguistics models by Nourani (1993, 1995).

From the practical view point the KR problems for first order logic formalisms as it is implied by Konolige's deductive view implies defining ways to apply links (Woods 1975). In Nourani-Lieberherr (1985) we showed how to define KR for automatic modelling with abstract objects for links in semantic nets (Schubert 1976). Hence the deductive view might benefit from our computational applications.

4. INTELLIGENT INTERFACES

4.1. Affective Computing and Multimedia Cognition

Picard (1999) assertions indicate not all modules is a designed AI system might pay attention to emotions, or to have emotional components. Some modules are useful rigid tools, and it is fine to keep them that way. However, there are situations where the human-machine interaction could be improved by having machines naturally adapt to their users.

Affective computing expands human-computer interaction by including emotional communication together with appropriate means of handling affective information. R. Picard's group addresses reducing user frustration; enabling comfortable communication of user emotion; developing infrastructure and applications to handle affective information; and building tools that help develop social-emotional skills. Since neurological studies indicate that the role of emotion in human cognition is essential; emotions are not a luxury. Instead, emotions play a critical role in rational decision-making, in perception, in human interaction, and in human intelligence.

These facts, combined with abilities computers are acquiring in expressing and recognizing affect, open new areas for research. The key issues in "affective computing," Piccard (1999a) computing that relates to, arises from, or deliberately influences emotions. New models are suggested for computer recognition of human emotion, and both theoretical and practical applications are described for learning, human-computer interaction, perceptual information retrieval, creative arts and entertainment, human health, and machine intelligence. Scientists have discovered many surprising roles played by human emotion—especially in cognitive processes such as perception, decision making, memory judgment, and more.

Human intelligence includes emotional intelligence, especially the ability to accurately recognize and express affective information. Picard suggests that affective intelligence, the communication and management of affective information in human/computer interaction, is a key link that is missing in telepresence environments and other technologies that mediate human-human communication. Picard-Cosier (1997) discusses new research in affective intelligence, and how it can impact upon and enhance the communication process, allowing the delivery of the more natural interaction that is critical for a true telepresence.

The techniques presented in the book are the bases to a haptic logic and multiagent cognition where central affecting computing questions might be addressed. The serious questions raised and addressed by Picard are as follows. Emotions might signal cognitive changes. Picard suggests that there is a small set of analog signaling mechanisms. Increasingly, neurologists, cognitive scientists, and psychologists are finding that human emotion biases memory formation, which, in turn, appears to bias decision making, creativity, planning, perception, judgment, mood-congruent memory retrieval, and a host of specific phenomena. The challenge remains for scientists to determine precisely how the many regulatory signaling cognitive functions and affective systems work their influences.

Picard argues that once we have duplicated in the machine all the important biasing, regulatory, motivational, behavioral, and other expressive phenomena associated with human emotions, and done so in a general, computable and flexible way, that we will have granted the machine emotions. Understanding and emulating human emotion might help in our efforts to understand intelligence. Are intelligent decisions based on emotions? If there is a Gestalt model for the world decided on, the answer might be affirmative. Furthermore, emotions are not independent of consciousness models (Nourani 1998, 1999, Chalmers 1999).

4.2. Knowledge-based Intelligent Interfaces

As we have seen thus far new advances in intelligent (knowledge-based) user interfaces that exploit multiple media text, graphics, maps—and multiple modalities—visual, auditory, gestural to facilitate human-computer interaction. The areas addressed are automated presentation design, intelligent multimedia interfaces, and architectural and theoretical issues. Mayberry (1997) is an edited volume on some of the original contributions in the area. There are three sections that address automated presentation design, intelligent multimedia interfaces, and architectural and theoretical issues. Automated Presentation Design: Intelligent Multimedia Presentation Systems:

Research and Principles; Planning Multimedia explanations using communicative acts, the automatic synthesis of multimodal Presentations; The Design of Illustrated Documents as a Planning Task; Automating the Generation of Coordinated Multimedia Explanations; Towards Coordinated Temporal Multimedia Presentations; Multimedia Explanations for Intelligent Training Systems. Intelligent Multimedia Interfaces are: The Application of Natural Language Models to Intelligent Multimedia. Enjoying the Combination of Natural Language Processing and Hypermedia for Information Exploration—An Approach to Hypermedia in Diagnostic Systems Integrating Simultaneous Input from Speech, Gaze, and Hand Gestures. The Architectural and Theoretical Issues: The Knowledge Underlying Multimedia Presentations; Using "Live Information" in a Multimedia Framework.

A Multilayered Empirical Approach to Multimodality: Towards Mixed Solutions of Natural Language and Graphical Interfaces; Modeling Issues in Multimedial Car-Driver Interaction. Multiagent multimedia navigation (Section 6.1) has been applied in our projects for spacecraft (Nourani 1996d) and terrain logics at IV-98, DaimlerBenz, Stuttgart and Nourani (1999d) on spatial navigation. Intelligent active multimedia databases are treated in our papers since 1998 and is amongst the areas on Chapter 7.

5. CONTEXT

A preliminary overview to context abstraction and meta-contextual reasoning is presented. Abstract computational linguistics (Nourani 1996b) with intelligent syntax, model theory and categories is presented in brief. Designated functions define agents, as in artificial intelligence agents, or represent languages with only abstract definition known at syntax. For example, a function F_i can be agent corresponding to a language Li. Li can in turn involve agent functions amongst its vocabulary.

Thus context might be defined at Li. An agent F_i might be as abstract as a functor defining functions and context with respect to a set and a linguistics model as we have defined. Generic diagrams for models are defined as yet a second order lift from context. The techniques to be presented have allowed us to define a computational linguistics and model theory for intelligent languages. Models for the languages are defined by our techniques in Nourani (1996a, 1987b) KR and its relation to context abstraction is defined in brief.

A computational linguistics with intelligent syntax and model theory is defined by Nourani (1996b, 1997a). Intelligent functions can represent agent functions, as artificial intelligence agents, or represent languages with definitions known at syntax. Since the languages represented by the agent functions can have arbitrary grammars not known to the signatures defined amongst the agent set, nondeterministic syntax computing is definable by the present linguistics theory. Form and context are definable by viewing computational linguistics by agent function sets. An agent FI might be as abstract as Functors defining functions and context with respect to a set and a linguistics model as we have defined.

To address the issues raised the role of context in KR and Natural Language systems, particularly in the process of reasoning is related to diagram functions defining relevant world knowledge for a particular context. The relevant world functions can transfer the axioms and relevant sentences for reasoning for a context. Further, by passing context around trees via intelligent syntax trees the locality burden is lifted from the deductive viewpoint. A formal computable theory can be defined based on the functions defining computable models for a context and the functions carrying context around.

For the VAS (Nourani 1997b) context foundations it is indeed possible to decrease the computational complexity of a formal system by the means of introducing context? Context localizes relevant worlds and specific computable functions define the world. Thus extraneous deductions are instant credits reducing complexity. The "what is context" question is reviewed in Section 3.3 from Section 4 we explore relations between contexts. Decontextualization is possible and might be necessary to address structural deductions. It might further be implied by paraconsistent logics (Nourani 1999a). Meta-contextual reasoning and a brief view to defining inter-context relations are introduced further on intelligent languages were presented in brief at Chapter 4.

Since the function symbols appearing might be invented by an activated agent without being defined in advance, intelligent syntax allows us to program with nondeterministic syntax. The parsing problems are quite challenging. Trees connect by message sequences hence carry parsing sequences with them. Thus the present computational linguistics theory is a start to programming with VAS (Nourani 1997b) and nondeterministic syntax. Other agent language projects are reported at Finin et al. (1997). We have defined intelligent context free grammars in Nourani (1997b) as follows. The mathematical specifics are treated in the foundational sections. We say that a language $\mathscr{L}$ is *intelligent context free*, abbreviated by ICF, iff $\mathscr{L}$ is intelligent and there is a context free grammar defining $\mathscr{L}$.

A preliminary parsing theory might be defined once we observe the correspondence between string functions and context. Let us define string intelligent functions. Functorial linguistic abstraction is where defining categories on the languages allows us to define lifts, for example, from context. Intelligent functions can represent agent functions, as in artificial intelligence agents, or represent languages with only abstract definition known at syntax. For example, a function F_i can be an agent corresponding to a language Li. Li can in turn involve agent functions amongst its vocabulary. Thus context might be defined at Li.

An agent F_i might be as abstract as a functor (Maclaine 1971, Goguen 1973) defining functions and context with respect to a set and a linguistics model as we have defined. Since the languages represented by the agent functions can have arbitrary grammars not known to the signatures defined amongst the agent set, nondeterministic syntax computing is definable by the present linguistics theory. This area is explored in Nourani (1996b, 1997a). Form and context are definable by viewing computational linguistics by agent function sets.

5.1. Models and Syntax

In the chapters and papers referred to we have presented computing with intelligent trees and objects, where intelligent tree rewriting as a formal algebraic and model-theoretic computing technique might be defined from the abstract syntax trees and language constructs. The generalized diagrams were defined by this author to encode the model-theoretic semantics of a language from its abstract syntax.

The techniques present language designs with linguistics constructs that make it easier to identify G-diagram models and define automatic implementations from abstract syntax.

There is a theory in principle for building models from syntax for first order logic. However, the computing enterprise requires more general techniques of model construction and extension, since it has to accommodate dynamically changing world descriptions and theories. The models to be defined are for complex computing phenomena for which we define.

5.2. Agent Linguistics

The linguistics abstraction techniques proposed allow us to lift from context to structures for analogical reasoning and proofs with free proof trees (Nourani 1995c). For example, the G-diagrams for models technique is applied at two levels for reasoning at meta-context. Models definable with G-diagrams allow free proof trees to be defined for meta-contextual reasoning with intelligent trees. The diagrams further define D⟨A,G⟩ categorical abstractions are defined for lifting from diagrams to categories for definable models. A third application for G-diagrams is for encoding situations—thus abstracting from Possible Worlds Context.

Categorical grammars are as close as computational linguistics has come to what we might want to refer by linguistics abstraction. The term categorical, however, is not quite in the same sense as in functorial linguistics abstraction (Nourani 1996b). There are recent techniques for structurally transforming abstract syntax by applying logical rules, for example functional composition, and abstraction. They are called categorical grammars (Lambek 1987, Konig 1990). These techniques have formed a basis for defining natural deduction like rules for grammars and proof techniques for abstract syntax trees by Koenig.

There are a number of references to the present author in the paper due to him having put forth the present area for computational logic only to show where it has been thus far.

Language designs with linguistics constructs are presented that make it easier to identify G-diagram models and define automatic programming from abstract syntax (Nourani 1993b, 1998d, 1995d).

5.3. Meta-Contextual Reasoning

What is context? Is context an inherent characteristic of natural language that ultimately decides the formal power of natural language? The abstract linguistics put forth by our linguistics abstraction have surprising implications. Utterance rich with abstractions, metaphors and string intelligent functions, i.e., functions and functors transcending context, is definable by a context free grammar. Abstract syntax and intelligent models are further presented in Chapter 6. Computing with intelligent trees (Nourani 1996a), G-diagrams for their models, and D⟨A,G⟩ categories are introduced in our mathematics projects published at ASL 1996-on, and applied to meta-contextual reasoning.

Meta-contextual reasoning is defined by lifting from syntax and clausal theories to proof theory with G-diagrams for intelligent trees D⟨A,G⟩ categories—categories for models definable by G-diagrams. Proof abstraction and planning with free proof trees (Nourani 1995c, Nourani-Hoppe 1994) are another technique for meta-contextual reasoning (Nourani 1999b). Relations between contexts can be defined by what context relevant functions are applied as to the context they correspond to and the context in which they appear. Intelligent signature functions transferring context around also define inter-context relations. A computer system can automatically infer the relation between some given set of contexts from the inter-context relevant functions.

5.4. KR, Models, and Context

Defining a category from the generalized diagram below is a second order lift from context. The G-diagram D⟨A,G⟩ defines a linguistics abstraction from content from which a linguistics model might be defined for reasoning. Abstract model theory as a second order lift is defined by a category D⟨A,G⟩. The D⟨A,G⟩ category is the category for models definable form D⟨A,G⟩. Knowledge representation has two significant roles: to define a model for the AI world, and to provide a basis for reasoning techniques to get at implicit knowledge.

Diagrams are the set of atomic and negated atomic sentences that are true in a model. Generalized diagrams are diagrams definable by a minimal set of functions such that everything else in the models closure can be inferred, by a minimal set of terms defining the model.

Thus providing a minimal characterisation of models, and a minimal set of atomic sentences on which all other atomic sentences depend. Our primary focus will be the relations amongst KR, AI worlds, and the computability of models. To keep the models which need to be considered small and to keep a problem tractable, such that the models could be computable and within our reach, are important goals (Nourani 1994). We show that we can apply G-diagram functions to localise reasoning to the worlds affected by some relevant functions to a specific reasoning aspect.

5.5. Diagrams and Incomplete Knowledge

In this section we extend the notion of generalized diagram (G-diagram) to include plausibility and nondeterminism for planning and for representation of possible worlds. An extended notion of G-diagram can encode possible worlds to capture the "maximally complete" idea and can be used for model revision and reconstruction. By assigning a plausibility ranking to formulas one can set a truth limit ordinal t as the truth threshold.

These notions of diagram are applied by way of example to planning such that the notions of computations with diagrams and free proof trees can be illustrated. A nondeterministic diagram is a diagram with indeterminate symbols instead of truth values for certain formulas. For example

$$\text{(nondeterministic diagram)} = \{p(c), q(c), p(f(t)), q(f(c)), q(f(f(c))), I\_q(f(s))\},$$

t is as defined by induction before; and $I\_q(f(s)) = I\_q$ for some indeterminate symbol I_q, for $\{s = t \text{ sub } n, n >= 2\}$

Formulas with plausibility ranking less than t would be assigned "T" and the other formulas would be assigned "F." Thus Nournai (1988, 1991) defined the notion of a plausible diagram, which can be constructed to define plausible models for revised theories. In practice, one may envision planning with plausible diagrams such that certain propositions are deliberately left indeterminate to allow flexibility in planning. In Nourani (1991) nondeterministic diagrams were defined by assigning an undefined "X" symbol to predicates in the diagram whose truth values are not known at each stage of planning.

Such extensions to the usual notion of diagram in model theory are put forth in Nourani (1988, 1991). That approach was one method of avoiding the computational complexity and computability problems of having complete diagrams. Truth maintenance and model revision can all be done by a simple reassignment to the diagram. The canonical model of the world is defined directly from the diagram. Generalized diagrams are shown to be an encoding for a minimal efficient knowledge representation technique applied to define relevant world models and implement reasoning trees. We have further shown how by defining predictive diagrams, partial deduction and abduction could be represented model-theoretically. We have also applied the techniques to proof abstraction and other related problems elsewhere.

6. MULTIAGENT VISUAL PLANNING

6.1. Intelligent Trees and Spatial KR

The visual field is represented by visual objects connected with agents carrying information amongst objects about the field, and carried onto intelligent trees for computation. Intelligent trees compute the spatial field information with the diagram functions. The trees defined have function names corresponding to computing agents. The computing agent functions have a specified module defining their functionality. The balloons are visual objects, the squares agents, the dotted lines the message paths.

Multiagent spatial vision techniques are introduced in Nourani (1998a,b). The duality for our problem solving paradigm (Nourani 1991a, 1995a,b) is generalized to be symmetric by the present paper to formulate double vision computing. The basic technique is that of viewing the world as many possible worlds with agents at each world that compliment one another in problem solving by cooperating. An asymmetric view of the application of this computing paradigm was presented by the author and the basic techniques were proposed for various AI systems (Nourani 1991a). The double vision computing paradigm with objects and agents might be depicted by the following figure. For computer vision (Winston 1975), the duality has obvious anthropomorphic parallels. The object co-object pairs and agents solve problems on boards by co-operating agents.

6.2. Virtual Trees and Planning with Models

This following sections presents techniques for automated planning and reasoning with virtual proof trees with applications to spatial computation. Generalized diagrams are applied to define models and implement plans. Free proofs are based on free Skolemization of diagrams and were put forth by Nourani (1995). A model theoretic view is defined to abductive computing by introducing abductive diagrams. The notion of a generalized diagram and the techniques of model building for AI world revision in earlier papers of this author is further developed to formulate free proof tree plan computation, model-theoretic planning. We also define GF diagrams and point out their application to algebraic plan computation. A sound free proof tree computation theorem and a Hilbert model theorem is presented for plan computation with free proof trees. The structural approach to the problem of planning and reasoning in artificial intelligence that was put forth it can be applied to AI problems in general based on the present formulation of AI worlds defined by GF diagrams. Thus a basis for AI computing with GF diagrams is presented. Boolean models are defined for diagram computations. An algorithm is presented for automated theorem proving and plan computation with virtual proof trees.

6.2.1. Multiagent Visual Planning

The co-operative problem solving paradigms have been applied ever since the AI methods put forth by Hays-Roth et al. (1985). See Nii (1986). The multiagent multi-board techniques due to Nourani (1995a), see next section.

The BID model has to be enhanced to be applicable to intelligent multimedia. Let us start with an example multi-board model where there multiagent computations based on many boards, where the boards corresponds to either virtual possible worlds or to alternate visual views to the world, or to the knowledge and active databases. The board notion is a generalization of the blackboard problem solving model (Hays-Roth 1985, Nii 1986).

The blackboard model consists of a global database called the blackboard and logically independent sources of knowledge called the knowledge sources. The knowledge sources respond opportunistically to the changes on the blackboard. Starting with a problem the blackboard model provides enough guidelines for sketching a solution. Agents can cooperate on a board with very specific engagement rules not to tangle the board nor the agents. The multiagent multi-board model, henceforth abbreviates as MB, is a virtual platform to an intelligent multimedia BID agent computing model. We are faced with designing a system consisting of the pair ⟨IM-BID,MB⟩, where IM-BID is a multiagent multimedia computing paradigm where the agents are based on the BID model. The agents with motivational attitudes model is based on some of the assumptions described as follows.

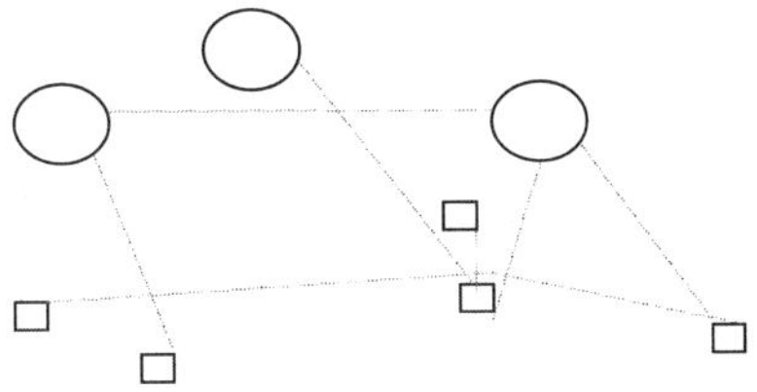

Figure 2. Agents and visual objects.

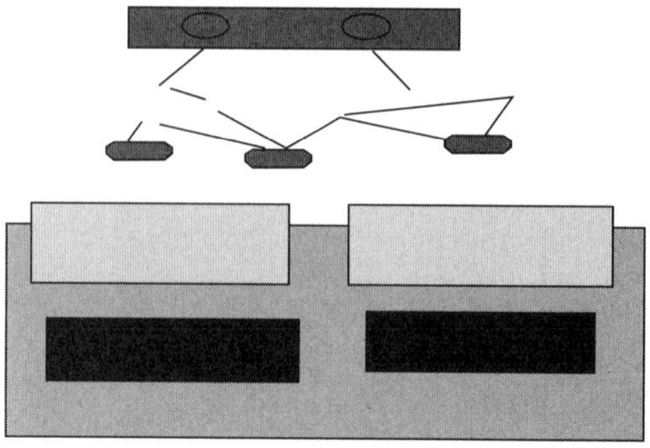

Figure 3. Multiagent multiboard computing platform.

Agents are assumed to have the extra property of rationality: they must be able to generate goals and act rationally to achieve them, namely planning, replanting, and plan execution. Moreover, an agent's activities are described using mentalistic notions usually applied to humans. To start with the way the mentalistic attitudes are modulated is not attained by the BID model. It takes the structural IM-BID to start it. The preceding chapters and sections on visual context and epistemics have brought forth the difficulties in tackling the area with a simple agent computing model.

The BID model does not imply that computer systems are believed to actually "have" beliefs and intentions, but that these notions are believed to be useful in modeling and specifying the behavior required to build effective multi-agent systems, for example Dennet (1996). The first BID assumption is that motivational attitudes, such as beliefs, desires, intentions and commitments are defined as reflective statements about the agent itself and about the agent in relation to other agents and the world. These reflective statements are modeled in DESIRE in a meta-language, which is order sorted predicate logic.

At BID the functional or logical relations between motivational attitudes and informational attitudes are expressed as meta-knowledge, which may be used to perform meta-reasoning resulting in further conclusions about motivational attitudes. If we were to plan with BID with intelligent multimedia the logical relations might have to be amongst worlds forming the attitudes and event combinations.

For example, in a simple instantiation of the BID model, beliefs can be inferred from meta-knowledge that any observed fact is a believed fact and that any fact communicated by a trustworthy agent is a believed fact. With IM_BID, the observed facts are believed facts only when a conjunction of certain worlds views and evens are in effect and physically logically visible to the windows in effect. Since planning with IM_BID is at times with the window visible agent groups, communicating, as two androids might, with facial gestures, for example Picard (1998).

In virtual or the "real-world" AI epistemics, we have to note what the positivists had told us some years ago: the apparent necessary facts might be only tautologies and might not amount to anything to the point at the specifics. Philosophers have been faced with challenges on the nature of absolute and the Kantian epistemtics (Kant 1990, Nourani 1999a) for years. It might all come to terms with empirical facts and possible worlds when it comes to real applications.

A second BID assumption is that information is classified according to its source: internal information, observation, communication, deduction, assumption making. Information is explicitly labeled with these sources. Both informational attitudes (such as beliefs) and motivational attitudes (such as desires) depend on these sources of information. Explicit representations of the dependencies between attitudes and their sources are used when update or revision is required.

A third assumption is that the dynamics of the processes involved are explicitly modeled. A fourth assumption is that the model presented below is generic, in the sense that the explicit meta-knowledge required to reason about motivational and informational attitudes has been left unspecified. To get specific models to a given application this knowledge has to be added. A fifth assumption is that intentions and commitments are defined with respect to both goals and plans. An agent accepts commitments towards himself as well as towards others (social commitments). For example, a model might be defined where a agent determines which goals it intends to fulfill, and commits to a selected subset of these goals. Similarly, an agent can determine which plans it intends to perform, and commits to a selected subset of these plans.

Most reasoning about beliefs, desires, and intentions can be modeled as an essential part of the reasoning an agent needs to perform to control its own processes. The task of belief determination requires explicit meta-reasoning to generate beliefs. Desire determination: Desires can refer to a (desired) state of affairs in the world (and the other agents), but also to (desired) actions to be performed.

Intention and commitment determination: Intended and committed goals and plans are determined by the component intention_and_commitment_determination. This component is decomposed into goal_determination and plan_determination. Each of these subcomponents first determines the intended goals and/or plans it wishes to pursue before committing to a specific goal and/or plan.

6.2.1.1. Agent Planning It shows that G-diagrams from the basis for minimal efficient knowledge representation, henceforth abbreviated by KR, paradigms. Areas touched upon are foundational and formal. Essential characteristics of visual representations, diagram understanding and interpretation, properties of animated and changing diagrams, diagram specification techniques, diagrammatic knowledge representation and inference, visual reasoning with diagrammatic programming languages, modelling interaction with diagrams, sound logical reasoning with diagrams, combination of diagrammatic knowledge and domain knowledge. We also show the applicability of G-diagram method for KR to partial deduction and abduction from Hoppe (1992), where we have defined abductive diagrams. Furthermore, we make a brief connection from Hoppe (1992) to KR for proof abstraction methods in AI (Nourani 1995, Mosetic-Holsbauer 1991), in the present paper, to show the applicability of the methods we are presenting with free proof trees (Nourani 1995b). Many examples are drawn from theories of AI to planning for robots to show the applicability of the techniques and theories proposed. World model revision is one of the difficult aspects of the nonmonotonic reasoning systems. Nourani (1993, 1988, 1991, 1999) present ways to get around it with generalized diagrams, where models are implicitly revised through dynamic changes to the diagrams from which the models are built.

Hence we show that KR by G-diagrams simplifies the world revision problem. The default consistency basis in Nourani (1991, 1993), as further applied to abductive reasoning in this paper, is referring to model-theoretic consistency. Generalized diagrams are used to build models with a minimal family of generalized Skolem functions. The minimal set of function symbols is functions with which a model can be built inductively. The functions can correspond to objects defining shapes and depicting pictures. We cannot formalize the real world. AI Worlds are relevant descriptions for problem solving in the real world parts we want models for G-diagrams can analogically model the semantics of a problem domain. The areas where we have applied G-diagrams include formal theories of DR (Nourani 1988, 1993, Nou-Hop 1994), Computational models of DR (Nourani 1991, Nourani 1995, Nou-Hop 1994, 1920). Application of DR in Nourani (1993, 1988, 1991, 1995), logic (Nourani 1995, Nou-Hop 1994), human-machine interfacing, visual languages have appeared in our papers. Applications to multimedia computing are presented in Nourani (1998a). The specific spatial areas addressed are cognitive structure of spatial knowledge, languages of spatial relations spatial and temporal reasoning, navigation in spatial environments, incomplete or imprecise spatial knowledge, presentation of spatial information virtual reality and robot navigation.

6.2.1.2. Diagrams and Models Here we present the notion of a predictive diagram and apply it for KR to provide a model-theoretic characterization for PD and related proof trees. A predictive diagram for a theory T is a diagram D(M), where M is a model for T, and for any formula q in M, either the function $f: q \rightarrow \{0, 1\}$ is defined, or there exists a formula p

in D(M), such that $T \cup \{p\}$ proves q; or that T proves q by minimal abduction. Hence it has a predictive function on incomplete knowledge. A generalized predictive diagram is a diagram with D(M) defined from specific functions. Abductive models could be minimally represented by a set of functions $\{f_1, \ldots, f_n\}$ that inductively define the model with a predictive diagram. The free trees (Nourani 1995) defined by the notion of provability implied by the definition, could consist of some extra Skolem functions $\{g_1, \ldots, g_l\}$,that appear at free trees. These extensions of PD to an abductive, nonmonotonic reasoning approach allow us to link them with the semantical approach to non-monotonic reasoning (Nourani 1991, 1995). By viewing PD from abductive diagrams we could define models for PD from abductive diagrams—thus a model theoretic formulation for PD emerges (Nou-Hop 1994, Nourani 1995b). We then define Hilbert models to handle the proof-model problems in Nou-Hop (1994).

6.3. Virtual Trees

In our projects present paper we also instantiate proof tree leaves with free Skolemized trees. Thus virtual trees are substituted for the leaves. In the present approach, as we shall further define, leaves could be virtual trees. By a virtual tree we intend a term made of constant symbols and Skolem functions terms A plan is a sequence of operations in the universe that could result in terms that instantiate the truth of the goal formulas in the universe. That's what goes on as far as the algebra of the model is concerned. It is a new view of planning prompted by our method of planning with GF-diagrams and free Skolemized trees.

It is a model-theoretic view. Proof-theoretically a plan is the sequence of proof steps (Fiks-Nils 1971) that yields the proof for the goal formula. The proof theoretic view is what the usual AI literature presents. The planning process at each stage can make use of GF-diagrams by taking the free interpretation, as tree-rewrite computations as in Nourani (1993), for example, of the possible proof trees that correspond to each goal satisfiability. The techniques we have applied are to make use of the free Skolemized proof trees in representing plans in terms of generalized Skolem functions. In planning with GF-diagrams that part of the plan that involves free Skolemized trees is carried along with the proof tree for a plan goal.

We can apply predictive diagram KR to compute queries and discover data knowledge from observed data and visual object images keyed with diagram functions. Model-based computing (Nourani 1995c, 1998b) which can be applied to automated data and knowledge engineering with keyed diagrams. Specific computations can be carried out with predictive diagrams (Nourani-Hoppe 1995). For cognition, planning, and learning the robot's mind a diagram grid can define state. The starting space applicable (Nourani 1991) project was meant for autonomous robots wandering at outer space. The designs in Nourani (1991, 1994, 1995) are ways for a robot to update its mind state based on what it encountered on its path. What the robot believes can be defined on a diagram grid. The degree to which a robot believes something is on the grid. It can get strengthened or weakened as a function of what the robot learns as progress is brought on. Robot's Mind State—The array grid entries are pointing to things to remember and the degree the robot believes them. The entry 15 is an item robot believes the most. Logically and theoretically the grid is minimally defined by the G-diagram functions. The gird model is a way to encode diagrammatic reasoning. Virtual trees and virtual tree planning is presented in further chapters.

6.4. Morph Gentzen Virtual Reality

The IM Morphed Computing Logic for computing for multimedia is new projects with important computing applications since (Nourani 1927, 1996a). The basic principles are a mathematical logic where a (Gentzen 1943) or natural deduction (Prawitz 1972) systems is defined by taking arbitrary structures and multimedia objects coded by diagram functions. Multimedia objects are viewed as syntactic objects defined by functions, to which the deductive system is applied. Thus we define a syntactic morphing to be a technique by which multimedia objects and hybrid pictures are morphed and, perhaps holographically, mapped via their defining functions to a new hybrid picture. Functorial topological structures can be defined without difficulty. The deduction rules are a Gentzen system augmented by Morphing, and Transmorphing.

The logical language has function names for hybrid pictures. The *MIM Morph Rule*—An object defined by the functional n-tuple $\langle f_1, \ldots, f_n \rangle$ can be morphed to an object defined by the functional n-tuple $\langle h(f_1), \ldots, h(f_n) \rangle$, provided h is a homomorphism of abstract signature structures (Nourani 1993c). The *MIM TransMorph Rules*—A set of rules whereby combining hybrid pictures $p_1, \ldots, p_n$ defines an Event $\{p_1, p_2, \ldots, p_n\}$ with a consequent hybrid picture p. Thus the combination is an impetus event. By transmorphing hybrid picture's corresponding functions a new hybrid picture is deduced. The techniques can be applied to arbitrary topological structures. The languages and MIM rules are applied to algebraic structures. The deductive theory is a Gentzen system in which hybrid pictures are named by parameterized functions; augmented by the MIM morph and transform rules. The model theory is defined from intelligent syntax languages (Nourani 1995d, 1996a). A computational logic for intelligent languages is presented in brief with a soundness and completeness theorem in Nourani (1996a). The idea is to do it at abstract models syntax trees without specifics for the shapes and topologies applied. We start with $\mathcal{L}\omega_1, \omega$ and further on might apply well-behaved infinitary languages. A soundness and completeness theorem has been put forth (Nourani 1997) stating the MIM soundness and completeness:

THEOREM 5.2 (NOURANI 1997). *MIM Logic is sound and complete.*

PROOF. Outline plain Morph Gentzen logical completeness has two proofs:

A—There is a direct proof which applies positive diagrams, and canonical models for $L\omega_1, \omega$ fragments as the authors' papers in mathematical logic since 1980.

B—There is a conventional proof route whereby we start with the completeness theorem for ordinary Gentzen systems. Form it we can add on the morph rules and carryout a proof based on what the morph rules preserve on models. Again intricate models are designed with positive diagrams.

6.5. Spatial Information and Morph Gentzen

Multiagent spatial vision techniques are introduced in the paper. The duality for our problem solving paradigm (Nourani 1991a, 1993c) is generalized to be symmetric by the present paper to formulate double vision computing. The basic technique is that of viewing the world as many possible worlds with agents at each world that compliment one another in problem solving by cooperating. The author presented an asymmetric view of the application of this computing paradigm and the basic techniques were proposed for various AI systems (Nourani 1991a). The double vision computing paradigm with objects and agents might be depicted by the following figure. For computer vision (Winston 1975), the duality has obvious anthropomorphic parallels. The object coobject pairs and agents solve problems on boards by cooperating agents. The cooperative problem solving paradigms have been applied ever since the AI methods put forth by Hays-Roth (1985). However, the multiagent multiboard techniques due to Nourani (1995a) Figure 1—lower boards depict the multiagent multiboard spatial vision design. The techniques to be presented are to be applied to Mobile multimedia. Communication and computation by multimedia visual-object languages can be programmed with IM with a simple syntax. The techniques to be presented are to be applied for (a) Precomputed video-object composition and combination for spatial morph Gentzen computing with visual objects (b) High speed visual spacecraft navigation by multiagent multimedia.

EXERCISES

1. A specific agent might have internal state set I, which the agent can distinguish its membership. The agent can transit from each internal state to another in a single step. With our multi-board models agent actions are based on I and board observations. There is an external state set S, modulated to a set T of distinguishable subsets from the observation viewpoint. A sensory function $s\colon \mathrm{S} \rightarrow \mathrm{T}$ maps each state to the partition

it belongs. Let A be a set of actions which can be performed by agents. A sensory function $s\colon S \to T$ maps each state to the partition it belongs. Let A be a set of actions which can be performed by agents. A function action can be defined to characterise an agent activity action: $T \to A$. There is also a memory update function mem: $I \times T \to I$. Define sensory functions and function action to characterise an agent carry on activities on a board modeling a world with two robots cooperating to put blocks on a lunar surface.

2. The BID model applies Task (de)composition. To model and specify (de)composition of tasks, we require knowledge of the following: a task hierarchy, information a task requires as input, information a task produces as a result of task performance meta-object relations between (sub)tasks Within a task hierarchy composed and primitive tasks are distinguished: in contrast to primitive tasks, composed tasks are tasks for which subtasks are identified. Subtasks, in turn, can be either composed or primitive. Tasks are directly related to components: composed tasks are specified as composed components and primitive tasks as primitive components. Design a preliminary agent implementation specified with task hierarchies applying a simple single-board blackboard model which can carry out agent tasking.
3. Applying a BID model where agent accepts commitments towards self as well as towards others (social commitments). For example, a model might be defined where a agent determines which goals it intends to fulfill, and commits to a selected subset of these goals. Similarly, an agent can determine which plans it intends to perform, and commits to a selected subset of these plans. Design a policy for selecting goals where the selection can be varied as goal satisfaction discovers new goals and plans which might take priority relative to the preselected initial goals. How do you maintain a knowledge base to accommodate your goal selection technique?
4. Most reasoning about beliefs, desires, and intentions can be modeled as an essential part of the reasoning an agent needs to perform to control its own processes. What changes have to be made to the BID design in Exercise 2 to design a IM_BID goal satisfaction.
5. Prove G-diagrams for models can encode possible worlds.
6. Design a preliminary basis for emotions sensed and sential computing where two androids are projected onto a board model to each. (a) Apply the IM basis above where the agents can bounce messages between the two boards to communicate as a basis to mutli-android affective computing. (b) What is a preliminary IM-BID model for it. (c) How does it compare to the planning problems at Section 4 and exercises there.
7. (a) Prove there is 1-1 correspondence between truth-sets and possible worlds. (b) Apply (a) to show compatibility can be obtained from diagrams.

REFERENCES

1. F. Badder, M. Buchheit, and B. Hollunder, *Artificial Intelligence* (1996).
2. J. Barwise, "The Situation in Logic-II: Conditional and Conditional Information." CSLI-85, Stanford, Ventura Hall, 1985a.
3. J. Barwise, "Notes on Situation Theory and Situation Semantics." CSLI Summer School, LICS, Stanford, 1985b.
4. M. A. Bratman, "Intentions, Plans, and Practical Reason." Harvard University Press, Cambridge, MA, 1987.
5. F. M. T. Brazier, B. Dunin-Keplicz, and N. R. Jennings, "Formal Specifications of Complex Reasoning Systems" (J. Treur and Th. Wetter Eds.), p. 282. E. Horwood, 1995.
6. F. M. T. Brazier, B. Dunin-Keplicz, N. R. Jennings, and J. Treur, "DESIRE: Modelling Multi-Agent Systems in a Compositional Formal Framework" (M. Huhns and M. Singh, Eds.), Vol. 1. International Journal of Cooperative Information Systems, 1997.
7. F. M. T. Brazier, J. Treur, N. J. E. Wijngaards, and M. Willems, "Temporal Semantics of Complex Reasoning Tasks" (B. R. Gaines and M. A. Musen, Eds.), Proc. of the 10th Banff Knowledge Acquisition for Knowledge-based Systems Workshop, KAW'95, SRDG Publications, Calgary, 1995.
8. F. M. T. Brazier, C. M. Jonker, and J. Treur, "Formalisation of a Cooperation Model Based on Joint Intentions," in Proc. of the ECAI'96 Workshop, 1996.
9. F. M. T. Brazier and J. Treur, "Compositional Modelling of Reflective Agents" (B. R. Gaines and M. A. Musen, Eds.). Proc. of the 10th Banff Knowledge, 1996.

10. Chris Bailey-Kellogg and Feng Zhao, "Spatial Reasoning Extracting and Reasoning with Spatial Aggregates," Vol. 24, No. 4. AI Magazine, 2004.
11. B. Chandrasekaran, Janice Galsgow, and N. H. Narayanan, Eds., " Diagrammatic reasoning 1995—Cognitive and Computational Perspectives." AAAI Press, 1995.
12. P. R. Cohen and H. J. Levesque, *Artificial Intelligence* 42, 213 (1990).
13. R. Cooper, J. Fox, J. Farrington, and T. Shallice, "The Completeness of Functional Calculus." A Systematic Methodology for Cognitive Modeling AI-85, pp. 3–44, 1996.
14. AI—Special Issue on Nomonotonic Logic, Vol. 13. 1980.
15. D. Dennett, "The Intentional Stance." MIT Press, Cambridge, MA, 1987.
16. E. Didday, "Knowledge Representation and Symbolic Data Analysis" (M Schader and W. Gaul, Eds.), Vol. F61. NATO AIS Series, Springer-Verlag, Berlin, 1990.
17. B. Dunin-Keplicz and J. Treur, "Compositional Formal Specification of Multi-agent Systems" (M. Wooldridge and N. R. Jennings, Eds.), Vol. 890, pp. 102–117. Intelligent Agents, Lecture Notes in Artificial Intelligence, Springer Verlag, Berlin, 1995.
18. D. L. Erman, B. Hays-Roth, F., V. L. Lesser, and D. R. Reddy, "The HEARSAY-II Speach Understanding System: Integrating Knowledeg to Resolve Uncertainty," Vol. 12, pp. 213–253. ACM Computing Survey, 1980.
19. R. E. Fikes and N. J. Nilsson, *Artificial Intelligence* 2, 189 (1971).
20. T. Finin, R. Fritzson, D. McKay, and R. McEntire, "KQML as an Agent Communications Language." Proceedings of the Third International Conference on Information and Knowledge Management, ACM Press, 1994.
21. J. A. Fodor, "The Language of Thought." T. W. Cromwell Company, New York, NY, 1975.
22. K. M. Ford, C. Glymour, and P. J. Hayes, "Android Epistemmology" (V. Lesser, Ed.). Formal Specification of Multi-Agent Systems: A Real-World Case. AAA/MIT Press, 1995.
23. P. Gardenfors, "Knowledge in Flux." MIT Press, 1988.
24. M. R. Genesereth and N. J. Nilsson, "Logical Foundations of Artificial Intelligence." Morgan-Kaufmann, 1987.
25. G. Gentzen, *Math. Ann.* 119, 140 (1943).
26. J. A. Goguen, J. W. Thatcher, E. G. Wagner, and J. B. Wright, "A Junction Between Computer Science and Category Theory." IBM Research Report RC4526, 1973.
27. B. Hays-Roth, *J. of Artificial Intelligence* 26, 251 (1985).
28. M. Heidegger, "Sein und Zeit" (J. S. R. Macquarrie and E. Robinson, Eds.), 8th ed. Harper & Row, New York, 1953.
29. M. Heidegger, Die Frage nach dem Ding, Max Niemeyer Verlag, Tubingen, 1962.
30. J. Hintikka, "Knowledge and Belief." Cornell University Press, Ithaca, NY, 1961.
31. J. R. Hobbs, M. Stickel, D. Appelt, and P. Martin, "Interpretation as Abduction," Tech. Note 499, SRI International, Menlo Park, CA, 1990.
32. Th. Hoppe, "On the Relationship between Partial Deduction and Abductive Reasoning," pp. 154–158. Proc. 10th ECAI Conference, John Wiley and Sons, Vienna, Austria, 1992.
33. J. Kagan, "The Nature of the Child." Basic, New York, 1984.
34. I. Kant, "Critique of Pure Reason" Translated by J. M. D Meiklejohn, 1990.
35. A. W. Kasznaic and A. C. Scott, "Towards a Science of Consciousness." The First Tucson Discussion and Debates, MIT Press, 1996.
36. D. Kinny, M. P. Georgeff, and A. S. Rao, "A Methodology and Technique for Systems of BID Agents" (W. van der Velde and J. W. Perram, Eds.), Vol. 1038. Agents Breaking Away, Proc. 7th European Workshop on Modelling Autonomous Agents in a Multi-Agent World, MAAMAW'96, Lecture Notes in AI, Springer Verlag, 1996.
37. S. Kleene, "Introduction to Metamathematics." 1952.
38. J. Koehler, "Planning From Second Principles," Vol. 890, pp. 1–39. AI 87, Lecture Notes in Artificial Intelligence, Springer Verlag, Berlin, 1996.
39. E. Koenig, "Parsing Categorical Grammars," Report 1.2.C, Espirit Basic Research Action 3175, Dynamic Interpretation of Natural Language (DYANA), 1990.
40. K. Konolige, "Belief and Incompleteness." Stanford CSLI-84-4, Ventura Hall, 1984.
41. S. A. Kripke, "Semantical Analysis of Modal Logics," Vol. 9, pp. 67–69. Zeitschrift fuer Mathematische Logik und Grundlagen der Mathematik, 1963.
42. G. Lakemeyer, *Artificial Intelligence* 84, 209 (1996).
43. J. Lambek, *American Mathematical Monthly* 65, 154 (1965).
44. M. Lewis, "Handbook of Emotions" (M. Lewis and J. Haviland, Eds.), Self-Conscious Emotions: Embarrassment, Pride, Shame, and Guilt, pp. 563–573. Guilford, New York, 1993.
45. S. Mac Lane, "Categories for the Working Mathematician." Springer-Verlag, NY, Heidelberg, Berlin, 1971.
46. M. T. Maybury, "Intelligent Multimedia Interfaces." MIT Press, 1997–1998.
47. A. Mehrabian, "Nonverbal Behavior." Aldine-Atherton, Chicago, 1972.
48. R. C. Moore, "Reasoning About Knowledge and Action," AI Center Technical Note 191, SRI International Menlo Park, California, 1980.
49. I. Mosetic and C. Holsbaur, "Extending Explanation Based Generalization by Abstraction Operators," Vol. 482. Machine Learning EWSL-91, LNAI, Springer-Verlag 1991.
50. J. P. Muller and M. Pischel, "Modeling Interactive Agents in Dynamic Environments," Proceedings 11th European Conference on AI. John Wiley and Sons, Amsterdam, The Netherlands, 1994.

51. P. H. Nii, "Blackboard Systems: The Blackboard Model of Problem Solving and the Evolution of Blackboard Architecture," pp. 38–53. The AI Magazine, 1986.
52. C. F. Nourani, "Equational Intensity, Initial Models, and AI Reasoning," Technical Report: A Conceptual Overview, in Proc. Sixth European Conference in Artificial Intelligence. Pisa, North-Holland, Italy, 1983.
53. C. F. Nourani, "Equational Intensity, Initial Models, and Reasoning in AI: A Conceptual Overview," Proc. Sixth European AI Conference, Pisa, North-Holland, Italy, 1984.
54. C. F. Nourani, "Nonmonotonicity, Plausibility and Structural Reasoning in Artificial Intelligence," Technical Report. CA, 1986.
55. C. F. Nourani, "Diagrams, Possible Worlds, and the Problem of Reasoning in Artificial Intelligence," Proc. Logic Colloquium, Journal of Symbolic Logic, Padova, Italy, 1987.
56. C. F. Nourani, "Planning and Plausible Reasoning in Artificial Intelligence, Diagrams, Planning, and Reasoning," Proc. Scandinavian Conference on Artificial Intelligence. IOS Press, Denmark, 1991.
57. C. F. Nourani, "Abstract Implementation Techniques for AI By Computing Agents," Technical Report: A Conceptual Overview, Proceedings SERF-93, Orlando, Florida, 1993a.
58. C. F. Nourani, "Automatic Models From Syntax," Proceedings XV Scandinavian Linguistics Conference. Oslo, Norway, 1993b.
59. C. F. Nourani, "Abstract Implementations By Computing Agents: A Conceptual Overview," Proc. SERF-93. Orlando, FL, 1993c.
60. C. F. Nourani, "GF-Diagrams For Models and Free Proof Trees." 1994.
61. C. F. Nourani, "Dynamic Epistemic Computing." Preliminary brief at the Summer Logic Colloquium, Claire-Mont Ferrand, France, 1994a.
62. C. F. Nourani, "Towards Computational Epistemology—A Forward," Proceedings Summer Logic Colloquium. Clermont-Ferrand, France, 1994b.
63. C. F. Nourani, "Free Proof Trees and Model-Theoretic Planning," AISB, Sheffield, England, 1995.
64. C. F. Nourani, "Double Vision Computing." IAS-4, Intelligent Autonomous Systems, Karlsruhe, Germany, 1995a.
65. C. F. Nourani, "Multiagent Robot Supervision," Learning Robots, Heraklion, 1995b.
66. C. F. Nourani, "Free Proof Trees and Model-Theoretic Planning." Proceedings Automated Reasoning, AISB, Sheffield, England, 1995c.
67. C. F. Nourani, "Language Dynamics and Syntactic Models," Proceedings Nordic and General Linguistics. Oslo University, Norway, 1995d.
68. C. F. Nourani, "Slalom Tree Computing—A Computing Theory For Artificial Intelligence," Vol. 9, No. 4. AI Communication, IOS Press, Amsterdam, 1996a.
69. C. F. Nourani, "Descriptive Computing," Summer Logic Colloquium, San Sebastian, Spain, 1996b.
70. C. F. Nourani, "Linguistics Abstraction," Brief Overview, Proceedings ICML96, International Conference on Mathematical Linguistics, Catalunya, Tarragona, Spain, 1996c.
71. C. F. Nourani, "Autonomous Multiagent Double Vision SpaceCrafts," AA99—Agent Autonomy Track, Seattle, WA, 1996d.
72. C. F. Nourani, "Computability, KR and Reducibility for Artificial Intelligence Problems," Vol. 4. No. 4. BSL, ASL, Toronto, 1997a.
73. C. F. Nourani, "MIM Logik." Summer Logic Colloquium, BSL Publications, ASL, 1997b.
74. C. F. Nourani, "Intelligent Languages—A Preliminary Syntactic Theory." Mathematical Foundations of Computer Science, 23rd International Symposium, Springer, Czech Republic, 1997c.
75. C. F. Nourani, "VAS-Versatile Abstract Syntax," 1997d.
76. C. F. Nourani, "A Parallel Functional Modular Language With Agents and Distributed Objects" Brief, IFL Wkshp Parallel Functional Languages, London, 1998a.
77. C. F. Nourani, "MIM Logik." Summer Logic Colloquium, Prague, 1998b.
78. C. F. Nourani, "Intelligent Trees, Thought Models, and Intelligent Discovery," Model-based Reasoning in Scientific Discovery (MBR'98). Pavia, Italy, 1998c.
79. C. F. Nourani, "Intelligent Multimedia," Intelligence and Multimedia Applications. Poster Announcement. Monash University, Australia, 1998d.
80. C. F. Nourani, "Visual Computational Linguistics and Visual Languages," Proceedings 34th International Colloquium on Linguistics, University of Mainz, Germany, 1998e.
81. C. F. Nourani, "Intelligent Multimedia—New Computing Technqiues and Its Applications." CSIT, Moscow, 1998f.
82. C. F. Nourani, "Syntax Trees, Intensional Models, and Modal Diagrams For Natural Language Models," Proceedings Uppsala Logic Colloquium. Uppsala University, Sweden, 1998g.
83. C. F. Nourani, "Morph Gentzen, KR, and SpatialWorld Models," 1998h.
84. C. F. Nourani, "Nonmontonic Model Culmination." Proceedings ASL, UCSD, La Jolla, 1999a.
85. C. F. Nourani, "Idealism, Illusion and Discovery," The International Conference on Mathematical Logic, Novosibirsk, Russia, 1999b.
86. C. F. Nourani, "Multiagent AI Design A New Emerging Software Engineering Trend," Vol. 12, pp. 37–42. Engineering Applications of AI-EAAI, Elsevier, 1999c.
87. C. F. Nourani, "Double Vision Spacecraft and Autonoumous Flight Control." AA99 Autonomy Control Track, Seatle, WA, 1999d.
88. C. F. Nourani, "Functorial Syntax and Paraconsistent Logics," BSL, ASL, New Orelans, 1999e.
89. C. F. Nourani, "Worlds, Models, and KM," Knowledge Representation and Philosphy, Luzern, 2002.

90. C. F. Nourani and T. Hoppe, "GF Diagrams, and Free Proof Trees." Technical Report, 1993.
91. C. F. Nourani and Th. Hoppe, "GF-Diagrams for Models and Free Proof Trees," Proceedings the Berlin Logic Colloquium, Homboldt Universty, 1994.
92. C. F. Nourani and K. J. Lieberherr, "Data Types, Direct Implementations, and KR," Proc. HICSS KR Track, Honollulu, Hawaii, 1985.
93. R. T. Picard, "Affective Computing, TR#321," MIT Media Lab. 1998.
94. R. W. Picard and G. Cosier, "Affective Intelligence—The Missing Link," Vol. 14, No. 4, pp. 56–71. BT Technology J., 1997.
95. D. Prawitz, "Natural Deduction: A Proof Theoretic Study." Almqvist and Wiksell, Stokholm, 1973.
96. Quine and W. Van Orman, "Word and Object," Harvard University Press, 1952.
97. A. S. Rao and M. P. Georgeff, "Modeling Rational Agents within a BID-Architecture" (R. Fikes and E. Sandewall, Eds.), pp. 473–484. Proceedings of the Second Conference on Knowledge Representation and Reasoning, Morgan Kaufmann, 1991.
98. B. Reeves and C. Nass, "The MediaEquation." Cambridge University Press, Cambridge, U.K., 1996.
99. L. K. Schubert, *Artificial Intelligence* 7, 163 (1976).
100. Y. Shoham, "Implementing the Intentional Stance" (R. Cummins and J. Pollock, Eds.), Philosophy and AI. MIT Press, Cambridge, MA, 1991.
101. Y. Shoham, *Artificial Intelligence* 60 (1993).
102. Y. Shoham and S. B. Cousins, "Logics of Mental Attitudes in AI: A Very Preliminary Survey" (G. Lakemeyer and B. Nebel, Eds.), pp. 296–309. Foundations of Knowledge Representation and Reasoning, Springer Verlag, 1994.
103. W. Sphon, "Ordinal Conditional Functions: A Dynamic Theory of Epistemic States" (W. L. Harper and B. Skyrms, Eds.), pp. 105–134. Causation, in Decision, Belief Change, and Statistics. Kluwer Academic Publishers, 1988.
104. W. van der Velde and J. W. Perram, Eds., "Agents Breaking Away," Vol. 1038. Proc. 7th European Workshop on Modelling Autonomous Agents in a Multi-Agent World, MAAMAW'96, Lecture Notes in AI, Springer Verlag, 1996.
105. D. Wilkins, *Artificial Intelligence* 22, 269 (1984).
106. M. Williams, "Explanation and Theory Base Transmutations," pp. 346–350. Proceedings 11th European Conference on AI, John Wiley and Sons Ltd., Amsterdam, 1994.
107. W. Woods, "Representation and Understanding" (D. G. Bobrow and A. Collins, Eds.). What is in a link? Academic Press, New York, 1975.
108. M. Wooldridge and N. R. Jennings, "Agent Theories, Architectures, and Languages: A Survey" (M. Wooldridge and N. R. Jennings, Eds.), pp. 51–92. Intelligent Agents, 1995.

CHAPTER 6

The Mathematical Foundations

CONTENTS

1. INTRODUCTION

Morph Gentzen and MIM—The IM Morphed Computing Logic for computing for multimedia are new projects with important computing applications since Nourani (1997). The basic principles are a mathematical logic where a Gentzen or natural deduction system is defined by taking arbitrary structures coded by diagram functions. The techniques can be applied to arbitrary topological structures. KR techniques with G-diagrams and applications to define computable models and relevant world reasoning are presented. G-diagrams are diagrams defined from a minimal set of function symbols that can inductively define a model. G-diagrams are applied to reasoning by model localized representations and a minimal efficient computable way to represent relevant knowledge for localized AI worlds. Diagrammatic reasoning is defined in terms of inferences directed by the G-diagrams for models. We show how computable AI world knowledge is representable. G-diagrams are applied towards KR from planning with nondeterminism and planning with free proof trees to partial deduction with predictive.

ISBN: 1-58883-037-3/$35.00

Intelligent Multimedia Computing Science
By Cyrus F. Nourani
Pages: 73–84

The application to proof abstraction is allude to and presented further in Nourani (1996a). A brief overview to a reasoning grid with diagrams is presented. MIM—the IM Morph Gentzen Logic for computing for multimedia is presented in formal terms with applications to diagrammatic reasoning. The basic principles are a mathematical logic where a Gentzen or natural deduction systems is defined by taking arbitrary structures and multimedia objects coded by diagram functions. By trans-morphing hybrid picture's corresponding functions a new hybrid picture is deduced. The techniques can be applied to arbitrary structures definable by infinitary languages, see for example, Shoenfield (1967). Multimedia objects are viewed as syntactic objects defined by functions, to which the deductive system is applied.

2. THE MORPH GENTZEN LOGIC

2.1. Intelligent Syntax and Models

By an intelligent language we intend a language with syntactic constructs that allow function symbols and corresponding objects, such that the function symbols are implemented by computing agents. Sentential logic is the standard formal language applied when defining basic models. The language $\mathcal{L}$ is a set of sentence symbol closed by finite application of negation and conjunction to sentence symbols. Once quantifier logical symbols are added to the language, the language of first order logic can be defined.

Structures for a language is defined as follows. For each function symbol in the language there is a function defined on A; and for each relation symbol in the language there is a relation defined on A. A Model $\mathcal{A}$ for $\mathcal{L}$ is a structure with a set A. There are structures defined for $\mathcal{L}$ such that for each constant symbol in the language there corresponds a constant in A. For the algebraic theories we are defining for intelligent tree computing in the forthcoming sections the language is defined from signatures as in the logical language is the language of many-sorted equational logic.

The signature defines the language $\mathcal{L}$ by specifying the function symbols' arities. The model is a structure defined on a many-sorted algebra consisting of S-indexed sets for S a set of sorts. A set of function symbols in the language, referred to by AF, is the set modelled in the computing world by AI agents. Thus the language $\mathcal{L}$ defined by the signature has designated function symbols called AF. For the specific foundations we refer to Nourani (1998a), for example. The AF function symbols define signatures that have specific message paths defined for carrying context around an otherwise context free abstract syntax. A set of function symbols in the language, referred to by AF, is the set of agents with nontrivial capability.

The computation is expressed by an abstract language that is capable of specifying modules, agents, and their communications. We have put together the AI concepts with syntactic constructs that could run on the tree computing theories we are presenting in brief. We have to define how the syntactic trees involving functions from the AF are to be represented by algebraic tree rewriting on trees. This is the subject of the next section, where free intelligent trees are defined. An important technical point is that for agents there are function names on trees.

DEFINITION 6.1. We say that a signature Σ is intelligent iff it has intelligent function symbols. We say that a language has intelligent syntax if the syntax is defined on an intelligent signature.

DEFINITION 6.2. A language $\mathcal{L}$ is said to be an intelligent language iff $\mathcal{L}$ is defined from an intelligent syntax.

Intelligent languages and signatures (Nourani 1998a) allow us to present computational theories with formulas on terms with intelligent function symbols. It is essential to the formulation of computations on intelligent trees and the notion of congruence that we define tree intelligence content. A reason is that there could be loss of tree intelligence content when tree rewriting because not all intelligent functions are required to be on mutual message exchanges. Theories are presented by axioms that define them and it is difficult to keep track of what equations not to apply when proving properties. What we have to define, however, is some computational formulation of intelligence content such that it applies to the present method of computability on trees. Once that formulation is presented, we could start decorating the trees with it and define computation on intelligent trees. It would be nice to view the problem from the standpoint of an example.

2.2. Basic Models, Completeness, and Morphisms

We have seen some basic notion on what languages and structures are in the preceding sections. Taking for granted first order logic and well-formed formulas, wffs (Kleene 1967, Enderton 1972) let us have a glimpse at some basic techniques and soundness-completeness properties for deductive systems.

DEFINITION 6.3. Let Γ be a set of wffs, φ a wff. Then Γ logically implies φ written $\Gamma \models \varphi$, iff for every structure A for the language and every function f: V $\rightarrow$ A such that A satisfies every member of Γ with f, A also satisfies φ with f.

DEFINITION 6.4. Let A and B be structures for the language. A homomorphism h of A into B is a function h: $|A| \rightarrow |B|$, where $|A|$ is the universe, i.e. the underlying set, for the structure $\mathfrak{R}$ such that

(c) For each n-place predicate letter p and each n-tuple $\langle a_1, \ldots, a_n \rangle$ of elements of $|A|$, $\langle a_1, \ldots, a_n \rangle$ is a n-place relation on A iff $\langle h(a_1), \ldots, h(a_n) \rangle$ is a n-place relation on B;
(d) For each n-place function symbol f and each such n-tuple, $h(f\text{A}(a_1, \ldots, a_n)) = f\text{B}(h(a_1), \ldots, h(a_n))$, where $f(h)$ is the function on the structure B defined for the function symbol f.

Homomorphisms are important to defining Morph Gentzen and morphings on multimedia computing. To start with the foundations to a sound deductive system we have to establish the premises for corresponding proofs and model where proofs can be carried out on propositions onto real structures. The basic first order logic soundness and completeness theorems are the following, where $\vdash$ denotes provability in a deductive system and $\models$ as in Definition 6.1.

The Soundness Theorem If $\Gamma \vdash \varphi$, then $\Gamma \models \varphi$.

Godel's Completeness (a) If $\Gamma \models \varphi$ then $\Gamma \vdash \varphi$; (b) Any constant set of formulas is satisfiable.

In the following sections we state the corresponding for Morph Gentzen and the unusual multimedia agent computing logic the author called MIM logic. As we had abbreviated in the preceding chapters, MIM and Morph Gentzen's syntactic morphing are techniques by which infinitary definable structures are homomorphically mapped via their defining functions to new structures. The basic pure mathematical basis to is a higher order deduction system we call Morph Gentzen designed with Gentzen sequents (Gentzen 1943) augmented by two rules Morphing, and Trans-morphing.

The Morph Rule—A structure definable by the functional n-tuple $\langle f_1, \ldots, f_n \rangle$ can be Morphed to a structures definable by the functional n-tuple $\langle h(f_1), \ldots, h(f_n) \rangle$, provided h is a homomorphism of abstract signature structures, see forthcoming sections and Nourani (1996a). *The TransMorph Rules*—A set of rules specific to structures applied to whereby combining structures $A_1, \ldots, A_n$ defines an Event $\{A_1, A_2, \ldots, A_n\}$ with a consequent structure B. Thus the combination is a sequent initiating event.

The deductive theory is a Gentzen system in which structures named by parameterized functions; augmented by the Morph and trans-Morph rules. The structures we apply the Morph logic to are definable by positive diagrams. The idea is to do it at abstract models syntax trees without specifics for the shapes and topologies applied. We have started with the infinitary language $\mathcal{L}_{\omega_1, \omega}$. Further on arbitrary well-behaved infinitary languages (Barwise 1968) might be applied. The important logical properties soundness and completeness can be proved for Morph Gentzen at the infinitary language $\mathcal{L}_{\omega_1, \omega}$ and the structures definable by such language (Nourani 1995b, 1996c).

THEOREM. *Soundness and Completeness—Morphed Gentzen logic is sound and complete.*

The proof applies some intricate and nontrivial model theory the author has published on infinitary languages with positive diagrams, e.g., Nourani (1995b). The mathematics and models with morphic preorders properties are further explored on our 1997 on projects. Proof outline Plain Morph Gentzen Logical Completeness has two proofs: (A) There is a direct proof which applies positive diagrams, and canonical models for $\mathcal{L}_{\omega_1,\omega}$ fragments as the authors' papers in mathematical logic since 1980. (B) There is a conventional proof route whereby we start with the completeness theorem for ordinary Gentzen systems. Form it we can add on the Morph rules and carryout a proof based on what the Morph rules preserve on models. Again intricate models are designed with positive diagrams. □

3. MORPHED GENTZEN VIRTUAL COMPUTING

To apply Morph Gentzen to multimedia morphing Morph Getzen had to be augmented with virtual trees with agent computation. Thus MIM a multiagent multimedia computing logic had been presented at Nourani (1997b). It is important to note that MIM is not simply anological reasoning. While anological computing might be built into the logic, the pictorial logic at MIM applies agents and proactive and reactive plans on events as the most important components. Morph Gentzen had been introduced as the purification to MIM to mathematicians subsequent to MIM, bit had been thought out before MIM got published. Agent computing with multimedia structures has been presented in the preceding chapters.

The mathematical bases to MIM applied computing on intelligent trees the author had invented a few years before MIM. Intelligent tree computing is the techniques and theory of computing with trees on signatures that bear agent functions on trees. Computability of initial models by tree rewriting are developed further for as a foundation for computing on trees to be applicable to intelligent tree computing (Nourani 1996a). The concepts of intelligent syntax and intelligent languages are presented.

The present applications are tree computing, agent tree rewriting is defined on intelligent trees by presenting the concepts and definition of algebraic tree intelligence content and mutual tree intelligence content within a forest. Models for intelligent theories emerge from the algebraic intelligent tree rewriting. Intelligent algebraic tree completion theorems and initial model rewrite theorems are put forth for intelligent trees. A soundness and completeness theorem is proved for intelligent tree rewriting as a formal model basis.

3.1. Computing on Trees

In order to present some motivation for the methods proposed certain model-theoretic concepts are reviewed and some new techniques are presented. The Henkin style proof for Godel's completeness theorem is implemented by defining a model directly from the syntax of theories. A model is defined by putting terms that are provably equal into equivalence classes, then defining a canonical structure on the equivalence classes (Henkin 1949). The computing enterprise requires more general techniques of model construction and extension, since it has to accommodate dynamically changing world descriptions and theories.

The models to be defined are for complex computing phenomena, for which we define generalized diagrams. The techniques in Nourani (1991) for model building as applied to the problem of AI reasoning allows us to build and extend models through diagrams. This required us to focus attention on generalized diagrams for models We had invented G-diagrams to build models with a minimal family of generalized Skolem functions.

The minimal set of function symbols is the set with which a model can be inductively defined. The models are initial and computable (Nourani 1996a). The G-diagram methods allowed us to formulate AI world descriptions, theories, and models in a minimal computable manner. Thus models and proofs for AI and computing problems can be characterized by models computable by a set of functions.

It allows us to program with objects and functions "running" on G-diagrams. To allude to our AI planning techniques as an example, the planning process at each stage can make use of G-diagrams with free Skolemized trees, by taking the free interpretation, as tree-rewrite computations for the possible proof trees that correspond to each goal satisfiability. Suppose there are some basic Skolem functions $f_1, \ldots, f_n$ that define a G-diagram. During planning or proof tree generation a set of Skolem functions $g_1, \ldots, g_n$ could be introduced. While defining such free proof trees, a set of congruence relations relates the g's to the f's. The proofs can make use of the tree congruence relations, or be carried out by tree rewriting.

The computing and reasoning enterprise require more general techniques of model construction and extension, since it has to accommodate dynamically changing world descriptions and theories. The techniques in Nourani (1991, 1995a, 1996a) for model building as applied to the problem of AI reasoning allows us to build and extend models through diagrams. This requires us to focus on the notion of generalized diagram. A technical example of algebraic models defined from syntax had appeared in defining initial algebras (Goguen et al. 1973, Thatcher et al. 1978) for equational theories of data types. In such direction for computing models of equational theories of computing problems are presented by a pair (Σ, E), where Σ is a signature (of many sorts, for a sort set S) and E a set of Σ-equations. Let $T\langle\Sigma\rangle$ be the free tree word algebra of signature Σ. The signature designates the functions names and their arties and co-arities with respect to the signature and the sorts. Chapter 12 has further precise mathematical definitions.

The quotient of $T\langle\Sigma\rangle$, the word algebra of signature Σ, with respect to the Σ-congruence relation generated by E, will be denoted by $T\langle\Sigma, E\rangle$, or $T\langle P\rangle$ for presentation P. $T\langle P\rangle$ is the "initial" model of the presentation P. The Σ-congruence relation will be denoted by $\equiv$P. One representation of T(P) which is nice in practice consists of an algebra of the canonical representations of the congruence classes, abbreviated by Σ-CTA. It is a special case of generalized standard models we had defined, see Nourani (1996a), for example. Some definitions are applied from our papers that allow us to define standard models of theories that are Σ-CTA's.

We apply generic diagrams to define canonical standard models in the same sense as set theory. The definitions are basic to sets and in defining induction for abstract recursion and inductive definitions. We had put forth variants of it with axiomatizations in our papers. The definitions were put forth by the first author for the computability with initial models. The canonical models are applied to multiagent computing during the last several years by Nourani (1996a, 1997a).

3.2. G-Diagrams for Initial Models

The G-diagrams for models are diagrams in which the elements of the structure are all represented by a minimal set of function symbols and constants, such that it is sufficient to define the truth of formulas only for the terms generated by the minimal set of functions and constant symbols. Such assignment implicitly defines the diagram. This allows us to define a canonical model of a theory in terms of a minimal family of function symbols. The minimal set of functions that define a G-diagram is the function set with which a standard model could be defined. Formal definition of diagrams are stated here, generalized to G-diagrams, and applied in the sections to follow.

Definition 6.5. Let M be a structure for a language $\mathscr{L}$, call a subset X of M a generating set for M if no proper substructure of M contains X, i.e., if M is the closure of $X \cup \{c(M)\colon c \text{ is a constant symbol of } \mathscr{L}\}$. An assignment of constants to M is a pair $\langle A, G\rangle$, where A is an infinite set of constant symbols in $\mathscr{L}$ and G: A $\rightarrow$ M, such that $\{G(a)\colon a \in A\}$ is a set of generators for M. Interpreting a by $g(a)$, every element of M is denoted by at least one closed term of $\mathscr{L}$(A). For a fixed assignment $\langle A, G\rangle$ of constants to M, the diagram of M, $D\langle A, G\rangle(M)$ is the set of basic (atomic and negated atomic) sentences of $\mathscr{L}$(A) true in M. (Note that $\mathscr{L}$(A) is $\mathscr{L}$ enriched with set A of constant symbols.)

Definition 6.6. A G-diagram for a structure M is a diagram $D\langle A, G\rangle$, such that the G in Definition 6.5 has a proper definition by specific functions.

For planning applications Fikes-Nilsson (1971), Nourani (1991) had presented some specific example functions. Thus initial standard models are models definable from their G-diagrams. Further practical and the theoretical characterization of models by their

G-diagrams are presented by this author in Nourani (1995a). It builds the basis for some forthcoming formulations that follow, and the tree computation theories that we had put forth. Method of constructing initial models by algebraic tree rewriting for the intelligent languages is to be developed from our approach in Nourani (1998a). We showed how initial algebras can be defined by subtree replacement and tree rewriting (Nourani 1996a). The G-diagram for the model is also defined from the same trees.

The examples of intelligent languages we could present have ⟨O, A, R⟩ triples as control structures. The A's have operations that also consist of agent message passing. The functions in AF are the agent functions capable of message passing. The O refers to the set of visual objects and R the relations defining the effect of A's on objects. Amongst the functions in AF, only some interact by message passing. The functions could affect objects in ways that affect the TID on the tree. There you are: the tree congruence definition is more complex for intelligent languages than those of ordinary syntax trees. Let us define tree intelligence content for the present formulation.

DEFINITION 6.7. Let a function f be a *string function*, iff there is no message passing or information exchange except onto the object that is at the range set for f, reading parameters visible at each object. Otherwise, f is said to be a splurge function. We refer to them by string and splurge functions when there is no ambiguity.

REMARK. Nullary functions are string functions.

DEFINITION 6.8. The tree intelligence degree, TID, is defined by induction on tree structures:

(i) the intelligence degree of a constant symbol function f is f;
(ii) for a string function f, and tree $f(t_1, \ldots, t_n)$ the TID is defined by $\cup$ TID $(t_i :: f)$, where $(t_i :: f)$ refers to a subtree of t_i visible to f;
(ii) for a *splurge* function f, TID is defined by $\cup$ TID $(f : t_i)$, where $f : t_i$ is the updated tree t_i since a single message exchange to t_i from f.

There are implicit mobile object computing principles at Definition 6.8, for example, the concept of a subtree being visible to a function, and of course, agents. The theorem below formalizes these points. Thus out of the forest of intelligent trees there appears an information theoretic rewrite theorems. Let us now define computing with intelligent equational theories. For further readings on the basics and current computing trends with algebraic trees and equations (Thatcher, Wagner and Wright 1978, Nourani 1979, Huet-Oppen 1980, Hussmann 1993, and Goguen et al. 1992) are starts. The above do not necessarily apply to intelligent object visual computing, but as a basis to start with equations on trees and abstract objects.

DEFINITION 6.9. A equational theory T of signature IΣ is an intelligent IΣ theory iff for every proof step involving tree rewriting, the TID is preserved. We state $\mathrm{T}\langle \mathrm{IST} \rangle \vdash t = t'$ when T is an IΣ theory.

DEFINITION 6.10. A equational theory T is intelligent, iff T has an intelligent signature IΣ, and axioms E, with IΣ its intelligent signature. A proof of $t = t'$ in an intelligent equational theory T is a finite sequence b of IΣ-equations ending in $t = t'$ such that if $q = q'$ is in b, then either $q = q'$ in E, or $q = q'$ is derived from 0 or more previous equations in E by one application of the rules of inference. Write $\mathrm{T}\langle \mathrm{IST} \rangle \vdash t = t'$ for "T p proves $t = t'$ by intelligent algebraic subtree replacement system."

By definition of such theories proofs only allow tree rewrites that preserve TID across a rule. These definitions have been applied to prove the theorems, set up the foundations for what could make intelligent tree rewriting TID, and define intelligent tree computation (Nourani 1996a). Thus the essence of intelligent trees will not be lost while agent tree computing.

Next, we define a computing agent function's intelligence content from the above definition. This is not as easy as it seems and it is a matter of the model of computation applied rather than a definition inherent to intelligent syntax. Let us present it as a function of intelligent syntax only, because we are to stay with abstract models and build models from abstract syntax. The definition depends on the properties of intelligent trees (Nourani 1996a) to be defined in the following section.

3.3. Intelligent Trees

Viewing the methods of computation on trees presented in the sections above we define intelligent trees here.

DEFINITION 6.11. A tree defined from an arbitrary signature Σ is intelligent iff there is at least one function symbol g in Σ such that g is a member of the set of intelligent functions AF, and g is a function symbol that appears on the tree.

DEFINITION 6.12. We define an intelligent Σ-equation, abbreviate by IΣ-equation, to be a Σ-equation on intelligent Σ-terms. A IΣ congruence is a Σ-congruence with the following conditions:

(i) the congruence preserves IΣ equations;
(ii) the congruence preserves computing agents intelligence content of Σ-trees.

Canonical models are definable with canonical sets C on the carriers with ⟨function, base-set⟩ pair by recursions such that C with a set of tree rewrite rules R represents T⟨IΣ, ~R⟩, where ~R is the set R of axioms for P viewed as IΣ-rewrite rules.

DEFINITION 6.13. The mutual intelligence content, MIC, of an intelligent function f, a member of the intelligent signature AF, is determined by the IΣ-congruence on T⟨AF⟩ relating the functions in AF. It is union of the TID over the trees that are a member of the congruence class of the free T⟨AF⟩ with respect to the IΣ-congruence defined on the T⟨Σ, w⟩, where w is the arity of f.

3.4. Intelligent Models

The author has proved how to obtain Canonical Intelligent Models (Nourani 1996). The theorems provide conditions for automatic implementation by intelligent tree rewriting to initial models for programming with objects.

THEOREM 6.1 (THE MIC THEOREM). *Let P be a presentation with intelligent signature IΣ for a computing theory T with intelligent syntax trees. Then T is*

(a) *A Sound logical theory iff every axiom or proof rule in T is TID preserving;*
(b) *A Complete logical theory iff there is a ⟨function,base-set⟩ pair defining a canonical set C and a G-diagram, such that C with R represents T⟨IΣ,~R⟩, where ~R is the set R of axioms for P viewed as IΣ-rewrite rules.*

PROOF. By definition of MIC, canonical rewrite theorems, completeness theorems for the first order logic, and completeness of induction for algebraic structures (Nourani 1996a). □

It is the logical foundations MIC theorem for intelligent syntax algebraic theories. At some further forthcoming research we might take a brief walk in the consequent fields and present new areas for computing. There are further MIC theorems that are relevant to computing and the model theory of computing with intelligent trees. Some of these applications appear in Double Vision Computing (Nourani 1995c).

4. VIRTUAL PLAN COMPUTING

4.1. Proof Trees

Virtual trees are computing trees where we instantiate proof tree leaves with free Skolemized trees. Thus virtual trees are substituted for the leaves. In the present approach, as we shall further define, leaves could be virtual trees. By a virtual tree we intend a term made of constant symbols and Skolem functions terms. A plan is a sequence of operations in the universe that could result in terms that instantiate the truth of the goal formulas in the universe. That's what goes on as far as the algebra of the model is concerned. It is a new view of planning prompted by our method of planning with GF-diagrams and free Skolemized trees.

It is a model-theoretic view. Proof-theoretically a plan is the sequence of proof steps (Fikes-Nilsson 1971) that yields the proof for the goal formula. The proof theoretic view is what the usual AI literature presents. The planning process at each stage can make use of GF-diagrams by taking the free interpretation, as tree-rewrite computations as in Nourani (1995a, 1996a), for example, of the possible proof trees. The techniques we have applied are to make use of the free Skolemized proof trees in representing plans in terms of generalized Skolem functions. In planning with G-diagrams that part of the plan that involves free Skolemized trees is carried along with the proof tree for a plan goal. Proofs can be abstracted by generalizing away from constants in the proof.

Thus, such a generalized proof can be defined by a whole class of minimal diagrams. Proof-theoretically we focus on SLDNF-resolution, which is a well-known deductive system. A SLDNF-proof can be considered as the unfolding of an AND/OR-tree, which is rooted in the formula to be proven, whose branches are determined by formulas of the theory, and whose leaves are determined by atomic formulas which are true in an AI world. In the present approach, as we shall further define, leaves could be free Skolemized trees.

By a free Skolemized tree we intend a term made of constant symbols and Skolem function terms. Allowing the proof-tree leaves get instantiated with non-atomic formulas, we get a more general notion of a proof, which is usually called "partial deduction" (Hoppe 1992), generalized to arbitrary leaves on diagrams to compute models at Nourani-Hoppe (1994). Partial deduction usually computes from a formula and a theory an existential quantified diagram.

The idea is that if the free proof tree is constructed then the plan has a model in which the goals are satisfied. The model is the initial model of the AI world for which the free Skolemized trees were constructed. Thus we had stated the Free Proof Tree Sound Computing Theorem since Nourani (1995a). We apply canonical models defined as follows to obtain initial models and plan goal satisfiability.

DEFINITION 6.14. Let (M, a)c in C be defined such that M is a structure for a language $\mathcal{L}$ and each constant c in C has the interpretation a in M. The mapping $c \to a_c$ is an assignment of C in M. We say that (M, a)c in C is canonical model for a presentation P on language $\mathcal{L}$, iff the assignment $c \to a$ maps C onto M, i.e. M = (a: c in C).

THEOREM 6.2. *For the free proof trees defined for a goal formula from the G-diagram there is an initial model satisfying the goal formulas. It is the initial model definable by the G-diagram.*

PROOF. (outline) In planning with GF-diagrams plan trees involving free Skolemized trees is carried along with the proof tree for a plan goal. The idea is that if the free proof tree is constructed then the plan has a model in which the goals are satisfied. There is analogy to SLD proofs. We can view on the one hand, SLD resolution type proofs on ground terms, where we go from $p(0)$ to $p(f(c))$; or from $p(f(c))$ to $p(f(g(c)))$. Whereas, while doing proofs with free Skolemized trees we are facing proofs of the form $p(g(\ldots))$ proves $p(f(g(\ldots)))$ and generalizations to $p(f(x))$ proves for all x, $p(f(x))$. Since the proof trees are either proving plan goals for formulas defined on the G-diagram, or are computing with Skolem functions defining the G-diagram, the model defined by the G-diagram applies and it is initial for the proofs. □

4.2. Virtual Plans

Generic diagrams, G-diagrams (e.g., Nourani 1995b) are used to build models with a minimal family of generalized Skolem functions. The minimal set of function symbols are those with which a model can be built inductively.

Proof-theoretically a plan is the sequence of proof steps that yields the proof for the goals, e.g., Genesereth-Nilsson (1987). In planning with G-diagrams that part of the plan that involves virtual Skolemnized trees is carried along with the proof tree for a plan goal.

Existentially quantified diagrams carry a main deficit—the Skolemized formulas are not characterized. Hilbert's epsilon symbol may be applied to solve this problem.

We then define Hilbert models to handle the proof-model problems further on. The idea is that if the virtual proof tree is constructed then the plan has a model in which the goals are satisfied. The model is the initial model of the AI world for which the virtual Skolemized trees were constructed. Thus we have the following Virtual Proof Tree Sound Computing Theorems. The following theorem from Nourani (1995a) establishes a correspondence from virtual proof trees to Hilbert models.

THEOREM 6.3. *The Hilbert's epsilon technique implies there is a model M for the set of formulas such that we can take an existentially quantified formula $w(X)$ and have it instantiated by a Skolem function which can answer the satisfiability question for the model.*

PROOF. We start with some Skolem functions to define the models. The equivalence defined by Hilbert's epsilon function is instantiated by the Skolem functions defining GF-diagrams. Thus we have Hilbert models for which the Skolem functions implicitly define membership to the set defining a model. While doing proofs with free Skolemized trees we are facing proofs of the form $p(g(\ldots))$ proves $p(f(g(\ldots)))$ and generalizations to $p(f(x))$ proves for all x, $p(f(x))$. Since the proof trees are either proving plan goals for formulas defined on the G-diagram, or are computing with Skolem functions defining the GF-diagram, the model defined by the GF-diagram applies for the proofs. □

5. MORPH GENTZEN, KR, AND MODELS

5.1. Knowledge Representation (KR)

This section is on the foundation for knowledge representation on generalized diagrams. G-diagrams presented in the preceding chapters form the basis for minimal efficient knowledge representation, henceforth abbreviated by KR, paradigms. Areas touched upon are: foundational and formal. Essential characteristics of visual representations, diagram understanding and interpretation, properties of animated and changing diagrams, diagram specification techniques, diagrammatic knowledge representation and inference, visual reasoning with diagrammatic programming languages, modelling interaction with diagrams, sound logical reasoning with diagrams, combination of diagrammatic knowledge and domain knowledge.

Furthermore, there is a brief connection form Nourani (1996a) to KR for proof abstraction methods in AI, in the present paper, to show the applicability of the methods we are presenting with free proof trees (Nourani 1995a). Thus we show that KR with G-diagrams for models simplifies the world revision problem (Nourani 1984).

5.2. Representing Relevant Knowledge

Knowledge representation has two significant roles: to define a model for the AI world, and to provide a basis for reasoning techniques to get at implicit knowledge. An ordinary diagram is the set of atomic and negated atomic sentences that are true in a model. Generalized diagrams are diagrams definable by a minimal set of functions such that everything else in the model closure can be inferred, by a minimal set of terms defining the model. Thus providing a minimal characterization of models, and a minimal set of atomic sentences on which all other atomic sentences depend.

We want to solve real world problems in AI. Obviously for automating problem solving, we need to represent the real world. Since we cannot represent all aspects of a real world problem, we need to restrict the representation to only the relevant aspects of the real world we are interested in. Let us call this subset of relevant real world aspects the *Relevant World* for a problem. Our primary focus will be the relations amongst KR, AI worlds, and the computability of models. The real world is infinite as are AI worlds at times. We might be interested to figure out in which AI worlds a theory or a sentence will be valid.

Furthermore, we might like to perform abstract inferences over equivalence classes of models (an issue of precomputation and reasoning). We have to be able to represent these ideas with computable formulations.

The usual Tarskian semantics allows us to use some symbols for denoting objects of the real world, to draw some inferences on those symbols and to obtain some statements that will then be true in the real world. Clearly, this is the basis for the symbolic computation paradigm of AI. However, the real world is complex, complicated and infinite. Thus we need to restrict any representation, so that it becomes computationally feasible. It is however possible, as we have abbreviated here and the preceding chapters, to define new symbolic computation paradigms for KR and based on G-diagrams, that have appealing computing properties. Hence, we focus usually during modeling on some subparts of the real world. What we do not know on generalized diagrams is defined in terms of generalized Skolem functions.

5.3. Further Diagrammatic Logics

The author had shown generalized diagrams to be an encoding for a minimal efficient knowledge representation technique applied to define relevant world models and implement reasoning trees. We have also applied the techniques to proof abstraction (Mostic-Holsbauer 1991) and other related problems appearing in the recent AI literature. We have defined various notions of generalized diagrams for KR applications to AI from planning and reasoning to implementing proof systems and proof abstraction (Nourani 1995a) to world revision problems. Generalized diagrams are promising KR technique. MIM computing is previewed as logic for multimedia and visual computing with diagrams.

The areas where the author has applied G-diagrams include formal theories of computational models of DR-digrammatical reasoning, the synergy between cognitive theories, formal theories, and computational models for cognitive theories. Application of DR with logic, human-machine interfacing, and visual languages have appeared in the author's papers referenced in the preceding chapters. Here are applications to new diagrammatic reasoning developments since our 1987 papers, for example, Diagrammatic (1995).

Most papers in such proceedings are on some computing basis to carrying images around. Starting with editor co-author Glasgow, where "important visual and spatial properties of an image," are examined there are areas presented in computational imagery by a Hamburg group-Habel et al. (1999) on panoramic depictions. Anological representations, abstract planning and perception, cognitive bases to visual interaction, and computational approaches to image understanding are amongst the areas. Problem solving paradigms with diagrams had received attention. Barwise-Etchemandy in Diagrammatic (1995) presented a heterogeneous logic, see Section 3, and Nourani (1980, 1984, 1991) to carry on logic and models with specific icons. It might be an area where specific computations are carried on at MIM and Morph Gentzen, specific instantiations might be applied with icons. What MIM and Morph Gentzen provide us with a abstract logic and model theory to carry on computation with physical image diagrams, designed with arbitrary structures and a complete agent planning basis. It applies functions to sleet arbitrary visual structures and pictures with a sound and complete logic. Specific icons and images are example instances of visual objects.

What the dynamic epistemic computing defines is not a situation logic in the Barwise sense. The situation and possible worlds concepts are the same Barwise. However, we define epistemics and computing on diagrams, with an explicit treatment for modalities. The correspondence of modalities to possible worlds and the containment of the possible worlds approach by the generalized diagrams implies that we can present a model-theoretic formulation for the dynamics of the possible worlds computing and for world descriptions for artificial intelligence applications. The number of situations exceeds the number of possible worlds. The possible worlds are the situations with no missing information and no contradictions. From the above definitions the mapping of terms and predicate models extend as in standard model theory. Next, a *compatible set of situations* is a set of situations with the same domain and the same mapping of function letters to functions. In other worlds, the situations in a compatible set of situations differ only on the truth conditions they assign to predicate letters.

DEFINITION 6.15. A situation consists of a nonempty set D, the domain of the situation, and two mappings: g, h. g is a mapping of function letters into functions over the domain as in standard model theory. h maps each predicate letter, p_n, to a function from D_n to a subset of $\{t, f\}$, to determine the truth value of atomic formulas as defined below. The logic has four truth values: the set of subsets of $\{t, f\}$. $\{\{t\}, \{f\}, \{t, f\}, 0\}$ the latter two corresponding to inconsistency, and lack of knowledge of whether it is true or false. □

DEFINITION 6.16. A G-diagram for a structure M is a model-theoretic diagram such that there is a proper diagram definition by a specific function set. □

REMARK. The minimal set of functions above are those by which a standard model could be defined by a monomorphic pair for the structure M. The dynamic of epistemic states as formulated by generalized diagrams is exactly what addresses the compatibility of situations. What the techniques lead us to is an algebra and model theory of epistemic states, as defined by generalized diagram of possible worlds. To decide compatibility of two situations we compare their generalized diagrams. Thus we have the following theorem.

THEOREM 6.4. *Two situations are compatible iff their corresponding generalized diagrams are compatible with respect to the Boolean structure of the set to which formulas are mapped by the function h, Definition* 6.15.

PROOF. The G-diagrams, Definition 6.16 encode possible worlds and since we can define a one-one correspondence between possible worlds and truth sets for situations compatibility is definable by the G-diagrams. □

One of the implications of the above towards cognition and computer vision is that of our notion of des Vielleicht Vorhandenen. It is within our mathematical expressive power to describe AI world situations. Thus we can represent an AI world and all the compatible generalized diagrams that can make "something" computationally tangible to base discovery on.

EXERCISES

1. Sound and complete logics are based on some well-behaved logical language properties. Starting with the space example in Chapter 5, what are some specifics the agents on a hybrid picture must satisfy to have a well-behaved intelligent agent language. Specific sound agents were presented at Skolem tree computing (Nourani 1996a). Can you define what a sound agent might be for the above example.
2. Define some Morph Gentzen sequences on the same example.
3. How might you define an intelligent context free language. Present an example of an intelligent context free language with two agents.
4. Prove Theorem 6.3.

REFERENCES

1. J. Barwsie, "Infintary Languages," Lecture Notes in Mathematics, Vol. 80. Springer-Verlag, Berlin.
2. "Diagrammatic Reasoning 1995—Cognitive and Computational Perspectives" (B. Chandraskekaran, Janice Glasow, and N. H. Narayaran, Eds.). AAAI Press, 1995.
3. H. B. Enderton, "A Mathematical Introduction to Logic." Academic Press, 1972.
4. R. E. Fikes and N. J. Nilsson, *Artificial Intelligence* 2, 189 (1971).
5. J. A. Fodor, "Imagistic Representation" (N. Blake, Ed.). In Imagery, The MIT Press, Cambridge, 1981.
6. M. R. Genesereth and N. J. Nilsson, "Logical Foundations of Artificial Intelligence." Morgan-Kaufmann, 1987.
7. G. Gentzen, *Math. Ann.* 119, 140 (1943).
8. J. A. Goguen, A. Stevens, K. Hobley, and H. Hiberdink, "A Metalogical Framework Based on Equational Logic" (C. A. R. Hoare and M. J. C. Gordon, Eds.), Vol. 339, pp. 69–86. Philosophical Transactions of the Royal Society, Prentice Hall, 1992.
9. J. A. Goguen, J. W. Thatcher, E. G. Wagner, and J. B. Wright, "A Junction Between Computer Science and Category Theory" (parts I and II), IBM Research Report RC4526, NY, 1973.
10. J. V. Guttag, The Specifications and Applications to Programming of Abstract Data Types, Ph.D Thesis. The University of Toronto, Canada, 1975.
11. Heinrich Hausmann, "Nonderminism in Algebraic Specifications and Algebraic Programs." Birkhauser, Boston-Basel-Berlin, 1993.
12. L. Henkin, *J. Symbolic Logic* 14, (1949).

13. Th. Hoppe, "On the Relationship Between Partial Deduction and Abductive Reasoning," pp. 154–158. Proc. 10th ECAI Conference, John Wiley and Sons, Vienna, Austria, 1992.
14. G. Huet and D. Oppen, "Formal Language Theory: Prospective and Open Problems" (R. V. Book, Ed.). Equations and Rewrite Rules, a Survey, SRI International Report, Academic Press, New York, 1980.
15. H. J. Keisler, "Model Theory for Infinitary Logic." North Holland, Amsterdam, 1971.
16. S. Kleene, "Mathematical Logic." John Wiley and Sons, 1967.
17. I. Mosetic and C. Holsbaur, "Extending Explanation Based Generalization by Abstraction Operators," Vol. 482. Machine Learning EWSL-91, Springer-Verlag, LNAI, 1991.
18. C. F. Nourani, "Abstract Specifications and Correctness of Implementations of Data Types and Algorithms." UCLA STOC Report, UMI, Ann Arbor, Michigan, 1979.
19. C. F. Nourani, "A Model-theoretic Basis to Specifications and Implementations of Abstract Data Types" (B. Robinet, Ed.). Programming Symposium, Lecture Notes in Computer Science, Springer-Verlag, Paris, 1980.
20. C. F. Nourani, "Equational Intensity, Initial Models, and AI Reasoning, Technical Report: A Conceptual Overview, in Proc. Sixth European Conference in Artificial Intelligence. Pisa, North-Holland, Italy, 1984.
21. C. F. Nourani, "Planning and Plausible Reasoning in AI," pp. 150–157. Proc. Scandinavian Conference in AI, IOS Press, Denmark, 1991.
22. C. F. Nourani, "Functorial Model Theory and Infinite Language Categories." ASL Bulletins, San Francisco, 1994.
23. C. F. Nourani, "Free Proof Trees, Model-Theoretic Planning and Nonderministic Algebraic Plan Computation with G-Diagrams." AISB, Sheffield, England, 1995a.
24. C. F. Nourani, "Functorial Model Theory." ASL, San Francisco, 1995b.
25. C. F. Nourani, "Double Vision Computing." Intelligent Autonomous Systems, Karlsruhe, Germany, 1995c.
26. C. F. Nourani, "Computable Functors and Generic Model Diagrams," Mathematical Foundations Topological Computing, Rohde Island, 1996.
27. C. F. Nourani, *The European AI J.* 4, (1996a).
28. C. F. Nourani, "AII Applications to Heterogenous Design." International Conference on Multiagent AI, Ronneby, Sweden, 1997a.
29. C. F. Nourani, "MIM Logik." ASL, Prague, 1997b.
30. C. F. Nourani, "Intelligent Languages—A Preliminary Syntactic Theory," p. 1450. Proc. 23rd International Mathematical Foundations Computer Science, Grammatics. Lecture Notes in Computer Science, Springer-Verlag, Brno, Czech Republic, 1998a.
31. C. F. Nourani and Th. Hoppe, "GF Diagrams, AI Models and Free Proof Trees." Berlin Logic Colloquium, Humboldt University, Berlin, 1994.
32. M. J. O'Donell, "Equational Logic as a Programming Language." The MIT Press, 1985.
33. D. Prawitz, "Natural Deduction: A Proof Theoretic Study." Almqvist and Wiksell, Stokholm, 1965.
34. G. Riekhiet and C. Habel, Eds., Mental Models in Discourse Processing and Reasoning, 259–297, Elsevier Science, 1999.
35. D. Scott, "Outline of a Mathematical Theory of Computation," pp. 169–176. Proc. 4th Annual Princeton Conference on Information Sciences and Systems, Oxford University Computing Lab., 1970.
36. J. W. Thatcher, E. G. Wagner, and J. B. Wright, "Data Type Specification: Parameterization and the Power of Specification Techniques," pp. 119–132. Proc. 10th Ann. Symposium on Theory of Computing (STOC), San Diego, CA, ACM, New York, 1978.

CHAPTER 7

Intelligent Visual Databases and Data Mines

CONTENTS

ISBN: 1-58883-037-3/$35.00

Intelligent Multimedia Computing Science
By Cyrus F. Nourani
Pages: 85–102

1. INTRODUCTION

The field of automated learning and discovery—often called data mining, machine learning, or advanced data analysis. The area has had obvious financial and organizational memory applications applied at times in our projects. E-Commerce, e-business, trust, trustworthiness, usability, human-computer interaction, cognitive ergonomics, user interface design, ease of use, ease-of-use, interaction design, and online marketing, are the user modeling issues the paper addresses. Financial companies have begun to analyze their customers' behavior in order to maximize the effectiveness of marketing efforts. There are routine applications to data discovery techniques with intelligence databases. Management process controls at times calls on warehoused data and relies on organizational memory to reach decision. Recent research has led to progress, both in the type methods that are available, and in the understanding of their characteristics. The broad topic of automated learning and discovery is inherently cross-disciplinary in nature. As there is increased reliance on visual data and active visual databases on presenting and storing organizational structures, via the internet and the WWW, the role of data discovery and intelligent multimedia active databases become essential. Knowledge management—KM and organizational memory OM are areas where model discovery with active intelligent databases applying predictive logic Nourani (1995b, 1999a, b) presented here applying intelligent multimedia databases (Nourani 2000). Knowledge Management (KM) is one of the key progress factors in organizations. In an organization, know-how may relate to problem solving expertise in functional disciplines, experiences of human resources, and project experiences in terms of project management issues, design technical issues and lessons learned. "Corporate memory" or "organizational memory" (OM) is regarded as the central prerequisite for IT supports of knowledge management. It is the means for knowledge conservation, distribution, and reuse. Identification and analysis of a company's knowledge-intensive work processes (e.g., product design or strategic planning), utilizing KM and OM are datadiscovery and datawarehousing intensive operations. A new computing area is defined with intelligent multimedia principles with business applications (Nourani 1999a). The area for which the paper provides a foundation for is where multimedia computing is bound to be applied to at dimensions and computing phenomena unimagined thus far, yet inevitable with the emerging technologies. The principles defined are new practical multimedia artificial intelligence business applications. Multimedia AI systems are proposed with new computing techniques defined. Multimedia objects, rules, and multimedia programming techniques are presented via a new language Nourani (1999d, 2000). The concept of hybrid picture is the start to define intelligent multimedia objects automatic hybrid picture transformation with a multimedia language in the author's projects. A preliminary mathematical basis to the Motph Gentzen computing logic is presented in Nourani (for example, 1999a, d, e; 2000). The foundations are a new computing logic with a model theory and formal system. Multimedia AI systems. Multimedia objects and rules are presented and shown in programming applications. Intelligent multimedia context defines the applications. Practical multimedia design is illustrated by pictorial examples. The project has applications to topics dealing with usability engineering, culture and design, international differences in software user training, case studies on international user interface design as in Del Galdo and Nielsen (1996). A practitioner for developing usable interfaces can apply intelligent multimedia to the practical designs (Browne 1994).

1.1. Where is Multimedia Data Mines?

The essential applications of computer technologies in business, economics, scientific, and personal interfaces is causing an exponential increase in digital data. Electronic database facilities are a standard technology. On specific applications and interfaces a next higher quality must be derived from the massive data storages. A next logical step would be the analysis of these data in order to derive useful information that is not easily accessible to humans. Data mining technology automatically derives new knowledge, new concepts, or knowledge structures from digital data. The starts were on numerical data, whereas multimedia applications call for methods that can play on documents, images, signals, or what combinations there might be. The sections to follow introduce the basic areas as well as the new foundations based on agent techniques and intelligent multimedia Morph Gentzen based DM. The emerging science and techniques are the basis for correlating and extracting information pieces from images and signals. The techniques are further applied to represent the corresponding knowledge in systems.

1.2. DM with Correlation Signatures

Standard examples are "query by example," where the user specifies an image, and the system attempts to retrieve the most similar images from the database.

To explore correlation signatures let us start with canonical correlation analysis, a classical multivariate method concerned with describing linear dependencies between sets of variables. Projection operators are used to illustrate the relations between canonical vectors and variates. A kernel feature space is presented spanned by objects. An exact solution for this kernel canonical correlation (KCCA) problem is derived from a geometric point of view. It shows that the expansion coefficients of the canonical vectors in their respective feature space can be found by linear CCA in the basis induced by kernel principal component analysis. The effect of mappings into higher dimensional feature spaces is considered critically since it simplifies the CCA problem in general.

Kernel methods attract a great deal of attention in the machine learning field of research initially due to the success of support vector machines. A common principle of these methods is to construct nonlinear variants of linear algorithms by substituting the linear inner product by kernel functions. Under certain conditions these kernel functions can be interpreted as representing the inner product of data objects implicitly mapped into a nonlinear related feature space (see for example Schölkopf and Smola (2002)).

Let $\mathbf{x}_i \in X$ $i = 1, \ldots, m$ denote input space objects and consider a feature space mapping $\phi\colon X \to F$ where the feature space F is an inner product space. The "kernel trick" is to calculate the inner product in F, $k(\mathbf{x}_i, \mathbf{x}_j) = \langle \phi(\mathbf{x}_i), \phi(\mathbf{x}_j) \rangle_F$, using a kernel function $k\colon X \times X \to R$ of input space objects while avoiding explicit mappings ϕ.

If an algorithm can be restated such that the data objects only appear in terms of inner products, one substitutes the linear dot product by such a kernel function. Though mappings will be used as an auxiliary concept during the construction of geometric algorithms, they never have to be constructed explicitly. The resulting kernel algorithm can be interpreted as running the original algorithm on the feature space mapped objects $\phi(\mathbf{x}_i)$.

This construction has been used to derive kernel variants of various methods originated in multivariate statistics. Prominent examples are kernel principal component analysis (Schölkopf et al. 1998), kernel discriminant analysis (Mika et al. 1999) and variants of chemometric regression methods like kernel principal component regression, kernel ridge regression and kernel partial least squares regression (Rosipal and Trejo 2001).

1.3. Learning and Multimedia Content

Kernel canonical correlation analysis is applied to learn a semantic representation of Web images and text. Examples are classification of images into categories, image retrieval based only on their content from a text query. The semantic space provides a common

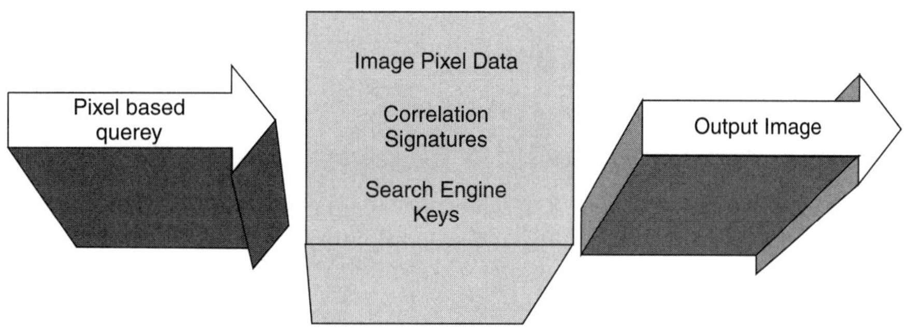

Figure 1. Intelligent multimedia datamining.

representation and enables a comparison between the text and image. A standard cross representation retrieval technique is the generalised vector space model. Since there are a vast amount of multimedia content in digital form on- and offline, it is important to process large amounts of mixed text image and video information. The combined image/text data processing has not been explored in the data analysis projects as much. Vinokourov et al. (2003) suggest a novel approach. Principal component analysis as applied (PCA) to text documents which can be considered to be generated by a mixture of a few independent sources, similarly in a way to newswire articles that come to a news agency from diverse sources. A landmark work in this area exploits nonnegative matrix factorisation (NMF), closely related to ICA, to analyse text and images content separately. Naive Bayes and extending it multinomial hierarchical asymmetric analysis (MASHA) also have proved to be quite successful approaches in text as an ICA-related alternative to LSI/PCA. However, in combination the semantic multimedia content analysis of different types of data (text and images) can bring more advantage than the analysis of each type of the data separately for, particularly, data categorisation problem. Kernel canonical correlation analysis (KCCA) to find correlations between web images and attached text. In this work the set of canonical directions, comprising the information about semantics of the text, is used not only to improve categorisation of the images but also to retrieve them according to text queries.

The algorithm presented in Vinokourov et al. (2003) shows connection between KCCA via incomplete Cholesky decomposition and Gram-Schmidt procedure. KCCA is applied to classify images and attached text and compare it with plain support vector machine (SVM) classification. Image retrieval using projection of text queries and images from a test collection onto the KCCA component. Feature space vectors are often found to give higher correlated features compared to the regularized kernel correlation solution. As in the linear case, the squared kernel canonical correlation coefficient describes the proportion of the sums of squares explained by the regression. Linear CCA also includes Fisher's linear discriminant analysis as a special case.

1.4. Learning Multimedia Content Semantics

In the following application the problem of learning semantics of multimedia content by combining image and text data is addressed. The synthesis is addressed by the kernel canonical correlation analysis described in Section 2. The learnt semantics is used for classification of images and their assigned text labels. The reported results show substantially better classification than for text or image taken alone that clearly demonstrates the ability of the method to learn the connection between the two different representations of multimedia Web content. The results also demonstrate an improvement over plain SVM classification in the combined vector space model.

The semantic representation is obtained by projecting data onto the semantic space spanned by vector solutions of the KCCA problem. For example, to process an unseen data point $\mathbf{x}$ we expand $\mathbf{x}$ into the vector representation for medium (image or text) space to get $\mathbf{ex}$ and then project it onto the canonical $\mathbf{F}$-correlation components, using the appropriate vector of the appropriate medium on the matrix with columns the first solutions of the KCCA problem for the given medium space sorted by eigenvalue in descending order.

2. THE VISUAL DYNAMICS

2.1. Why Multimedia Databases?

Defining compatibility and visual effects relations, allows objects to be selected and applied to design and compare customized views. Multimedia programming is combined with visual multiagent objects to define specific visual compatibility for customized active databases. Active databases deploy certain computing which lend themselves naturally to the Intelligent Multimedia principles Nourani (1999d). The concept of an active objects are embedded by intelligent objects and "events" are embedded by the computing defined by IM as a basic principle an embedded by intelligent trees, intelligent objects, and hybrid pictures. The characteristics of an active DMBS, or ADMBS, see for example Bailey (1995), supports definition and management of ECA-rules, e.g. event, condition, and action. Hence an ADMBS must have means to define ECA's. An ADBMS must support rule management and rule-base updates. It must carry out actions and evaluate conditions. An ADBMS must represent information in ECA-rules in terms of its data models. The IM computing paradigm provides a basis for designing multimedia ADBMS's. The IM computing paradigm allows the design for multimedia ADBMS Nourani (2000) to apply agent computing to ADBMS, to base an ADBMS on multimedia agent computing, and to carryout metalevel reasoning and KB with multimedia intelligent objects.

2.2. KR and Diagrams for Relevant World Models

Let us brief on knowledge representation with G-diagram models Nourani (1995b, 1999a) and applications to define computable models and relevant world reasoning. G-diagrams are diagrams defined from a minimal set of function symbols that can inductively define a model. G-diagrams are applied to relevance reasoning by model localized representations and a minimal efficient computable way to represent relevant knowledge for localized AI worlds. We show how computable AI world knowledge is representable. G-diagrams are applied towards KR from planning with nondeterminism and planning with free proof trees to partial deduction with predictive diagrams applied in Nourani (1999a, b, d). The applications to proof abstraction and explanation-based generalization by abstract functions are alluded to in Nourani (1995b). A brief overview to a reasoning grid with diagrams is presented in the above referenced papers. Generalized diagrams are used to build models with a minimal family of generalized Skolem functions. The minimal sets of function symbols are functions with which a model can be built inductively. The functions can correspond to objects defining shapes and depicting pictures. We cannot formalize the real world, however, the relevant descriptions for problem solving can be specified. Knowledge representation has two significant roles: to define a model for the AI world, and to provide a basis for reasoning techniques to get at implicit knowledge. An ordinary diagram is the set of atomic and negated atomic sentences that are true in a model. Generalized diagrams are diagrams definable by a minimal set of functions such that everything else in the model's closure can be inferred, by a minimal set of terms defining the model. Thus providing a minimal characterization of models, and a minimal set of atomic sentences on which all other atomic sentences depend. To prove Godel's completeness theorem, Henkin defined a model directly from the syntax of the given theory. This structure is obtained by putting terms that are provably equal into equivalence classes, then defining a free structure on the equivalence classes.

The computing enterprise requires more general techniques of model construction and extension, since it has to accommodate dynamically changing world descriptions and theories. Let us define a simple language $\mathscr{L} = \langle\{\text{tweety}\}, \{a\}, \{\text{bird}\}$, predicate letters at first order logic—FOL$\rangle$. A model may consist of {bird(tweety), $\neg$ penguin(tweety) $\rightarrow$ (bird(tweety), bird(tweety) v(bird(tweety),...}, others may consist of $\{p(a), \neg p(a) \rightarrow p(a), p(x), p(a)vp(x)\ vp(y), \ldots\}$. Because we can apply arbitrary interpretation functions for mapping language constructs into worldviews, the number of models for a language is infinite. Although this makes perfect sense from a theoretical and logical point of view, from a practical point of view, this notion of model is too general for our applications. Since for AI we want models that could be computed effectively and efficiently. Thus, it is useful to restrict the types of models that we define for real world applications. Primarily, we are interested in models with computable properties definable from the theory. The generic diagram, G-diagram for models Nourani (1991, 1995b, 2001) is a diagram in which the elements of the structure are all represented by a minimal family of function symbols and constants. Thus it is sufficient to define the truth of formulas only for the terms generated by the minimal family of functions and constant symbols. Such assignment implicitly defines the diagram. This allows us to define a canonical model of a theory in terms of a minimal function set. The following sections apply KR with visual business objects and Section 5 presents the applications to heterogeneous computing and model discovery on business data.

3. A VISUAL COMPUTING LOGIC

A problem-solving paradigm is presented in the double vision computing paper (Nourani and Loo 1999). The basic technique to be applied is viewing the televised scene combined with the scripts as many possible worlds. Agents at each world that compliment one another to portray a stage by cooperating. The AI techniques can be applied to define interactions amongst personality and view descriptions. Object co-object pairs and agents solve problems on boards by cooperating agents from the pair without splurges across the pairs. The term splurge has a technical definition for object level computing presented in Nourani (1995a). Computing by agents might apply the same sort of cooperative problem solving methods. The intelligent multimedia computing paradigm can define multiagent computing with multimedia objects and carry on artificial intelligence computing on boards. The IM hybrid multimedia programming techniques have a computing logic is mathematical logic where a Gentzen (1943) or natural deduction Prawitz (1965) systems is defined by taking multimedia objects coded by diagram functions. By transforming hybrid picture's corresponding functions a new hybrid picture is deduced. Multimedia objects are viewed as syntactic objects defined by functions, to which the deductive system is applied. Thus we define a syntactic morphing to be a technique by which multimedia objects and hybrid pictures are homomorphically mapped via their defining functions to a new hybrid picture. The deduction rules are a Gentzen system augmented by Morphing, and trans-Morphing. The logical language has function names for hybrid pictures. The Morph Rule—An object defined by the functional n-tuple $\langle f_1, \ldots, f_n \rangle$ can be morphed to an object defined by the functional n-tuple $\langle h(f_1), \ldots, h(f_n) \rangle$, provided h is a homorphism of intelligent objects as abstract algebras, and $h(f)$ is the corresponding structure to what f selects, at the morphed world. The trans-Morph Rules—A set of rules whereby combining hybrid pictures $p_1, \ldots, p_n$ defines an event $\{p_1, p_2, \ldots, p_n\}$ with a consequent hybrid picture p. Thus the combination is a Morph sequencing initiating event. The deductive theory is a Gentzen system in which hybrid pictures are named by parameterized functions; augmented by the MIM Morph and transmorph rules. The formal AI and mathematics appear in the first author's mathematical logic publications since 1997 and applied to business areas at Nourani (1999a, d) for example. The intelligent syntax languages are applied with Morph Gentzen (1943).

3.1. Intelligent Models

A computational logic for intelligent languages is presented in brief with a soundness and completeness theorem in Nourani (1999a) from Nourani (1994) and applied to intelligent business object computing. A brief overview to context abstraction shows how context free and context sensitive properties might be defined with agents. Agents are in the sense defined in Nourani (1995a) and the AI foundations Genesereth (1987). A set of function symbols in the language, referred to by agent function set, is the set of function symbols that are modeled in the computing world by AI agents. The idea is to characterize computation with abstract syntax trees without grammar specifics. As an example, suppose you are told there is an academic department with a faculty member which is Superman, and two faculty members which are Swedish speaking, and three which do not talk to anybody not in their expertise area. Without telling you anything else about what they do, abstract syntax properties can be defined. When the signature's specific agent functions are defined, it implies the signature has defined message paths for them. From the signature intelligent models are defined for abstract syntax trees. The computation is expressed by an abstract language that is capable of specifying modules, agents, and their communications. The implementing agents, their corresponding objects, and their message passing actions can also be presented by the two-level abstract syntax. The agents are represented by function names that appear on the free syntax trees of implementing trees. The trees defined by the present approach have function names corresponding to computing agents. The computing agent functions have a specified module defining their functionality. A signature defines the language tree compositionality degree and defines the abstract syntax. The definitions prompted a computational linguistics and model theory for intelligent languages. Models for the languages are defined by our techniques in Nourani (1995a, 2001).

3.2. Relevant KR and Visual Business Objects

Knowledge representation has two significant roles: to define a model for the AI world, and to provide a basis for reasoning techniques to get at implicit knowledge. An ordinary diagram is the set of atomic and negated atomic sentences that are true in a model. Generalized diagrams are diagrams definable by a minimal set of functions such that everything else in the model's closure can be inferred, by a minimal set of terms defining the model. Thus providing a minimal characterization of models, and a minimal set of atomic sentences on which all other atomic sentences depend. We want to solve real world problems in AI. Obviously for automating problem solving, we need to represent the real world. Since we cannot represent all aspects of a real world problem, we need to restrict the representation to only the relevant aspects of the real world we are interested in. Let us call this subset of relevant real world aspects the relevant world for a problem. AI approaches to problem solving represent the knowledge usually in some kind of first-order language, consisting of at least constants, functions and predicate symbols. Our primary focus will be the relations amongst KR, AI worlds, and the computability of models. Truth is a notion that can have dynamic properties. To keep the models which need to be considered small and to keep a problem tractable, we have to get a grip on a minimal set of functions to define computable models with. The selector functions are applied to create compound business objects.

3.3. Learning from Mutimedia Data

Many data sets contain more than just a single type of data. The existing algorithms can usually only cope with a single type data. How can we design methods that can take on multimedia data from multiple modalities? Are we to apply separate learning algorithms to each data modality, and combine their results, or are there to be algorithms that can handle multimedia multiple modalities on a feature-level. Learning casual relationships amongst visual stored data is another important area, which can benefit from our project. Most existing learning algorithms detect only correlations, but are unable to model causality and hence fail to predict the effect of external controls. Visualization and interactive discovery data mining is a process which involves automated data analysis and control decisions by an expert of the domain. For example, patterns in many large-scale business system databases might be discovered interactively, by a human expert looking at the data, as it is done with medical data. Data visualization is specifically difficult when data is high-dimensional, specifically when it involves non-numerical data such as text. The projects might be a basis for designing interactive business tools.

3.4. Computable World Models

The techniques in Nourani (1991, 1995b, 2001) for model building as applied to the problem of AI reasoning allows us to build and extend models by diagrams. This requires us to define the notion of generalized or generic diagram. The G-diagrams are used to build models with a minimal family of generalized Skolem functions. The minimal sets of function symbols are those with which a model can be built inductively. We focus our attention on such models, since they are computable (Bailey et al. 1995, Nourani 1991). The G-diagram methods applied and further developed here, allows us to formulate AI world descriptions, theories, and models in a minimal computable manner. It further allows us to view the world from only the relevant functions. Thus models and proofs for the specified problems can be characterized by models computable by a set of functions. The G-diagram functions can define IM objects and be applied with the Morph Gentzen logic.

4. HETEROGENEOUS COMPUTING MODELS

4.1. Heterogeneous Computing

In Nourani (1999) we present new techniques for design by software agents and new concepts entitled abstract intelligent implementation of AI systems (AII). Objects, message passing actions, and implementing agents are defined by syntactic constructs, with agents appearing as functions. The techniques have been applied to design intelligent business objects in Nourani (1999). AII techniques have been applied to heterogeneous KB design and implementation. The application areas include support for highly responsive planning. AII techniques are due to be an area of crucial importance as they are applied gradually to the real problems. The applied fields are, for example, intelligent business systems, aerospace, AI for robots, and multimedia.

4.2. KR, Knowledge Base, and Visual Model Discovery

Model diagrams allow us to characterize incomplete KR. To key into the incomplete knowledge base we apply generalized predictive diagrams whereby specified diagram functions a search engine can select onto localized data fields. The predictive model diagrams Nourani (1995b, 1999a) could be minimally represented by the set of functions $\{f_1, \ldots, f_n\}$ that inductively define the model. Data discovery from KR on diagrams might be viewed as satisfying a goal by getting at relevant data which instantiates a goal. The goal formula states what relevant data is sought. We propose methods that can be applied to planning Nourani (1991) with diagrams to implement discovery planning. In planning with G-diagrams that part of the plan that involves free Skolemized trees is carried along with the proof tree for a plan goal. Computing with diagram functions allows us to key to active visual databases with agents.

4.3. Prediction and Discovery

Minimal prediction is an artificial intelligence technique defined since the author's model-theoretic planning project. It is a cumulative nonmontonic approximation attained with completing model diagrams on what might be true in a model or knowledge base. A predictive diagram for a theory T is a diagram D(M), where M is a model for T, and for any formula q in M, either the function $f\colon q \to \{0, 1\}$ is defined, or there exists a formula p in D(M), such that $T \cup \{p\}$ proves q; or that T proves q by minimal prediction. A generalized predictive diagram, is a predictive diagram with D(M) defined from a minimal set of functions. The predictive diagram could be minimally represented by a set of functions $\{f_1, \ldots, f_n\}$ that inductively define the model. The free trees we had defined by the notion of provability implied by the definition, could consist of some extra Skolem functions $\{g_1, \ldots, g_l\}$ that appear at free trees. The f terms and g terms, tree congruences, and predictive diagrams then characterize partial deduction with free trees. The predictive diagrams are applied to discover models to the intelligent game trees. Prediction is applied to plan goal satisfiability and can be combined with plausabilities Nourani (1991), probabilities, and fuzzy logic to obtain, for example, confidence intervals. The first author has applied minimal prediction to simply encode knowledge with model diagrams to carry on automated deduction as Loveland and Poole's automated deduction system intended. Modeling with virtual tree planning Nourani (1999c) is applied where uncertainty, including effector and sensor uncertainty, are relegated to agents, where competitive learning on game trees determines a confidence interval. The incomplete knowledge modelling is treated with KR on predictive model diagrams. Model discovery at KB's are with specific techniques defined for trees. Model diagrams allow us to model-theoretically characterize incomplete KR. To key into the incomplete knowledge base we apply generalized predictive diagrams whereby specified diagram functions a search engine can select onto localized data fields. The predictive model diagrams could be minimally represented by the set of functions $\{f_1, \ldots, f_n\}$ that inductively define the model.

4.4. TAIM and Discovery from Warehoused Data

The morphing Transformer Active Intelligent Database designs outline in the paper is abbreviated as TAIM[1] hereforth. The idea is that if the free proof tree is constructed then the plan has a model in which the goals are satisfied. The model is the initial model of the AI world for which the free Skolemized trees were constructed. Partial deductions in this

[1] TAIM is the tradename for the active intelligent database design and is the technical property of the authors computing business. Applications must be with permission and venture partnerships.

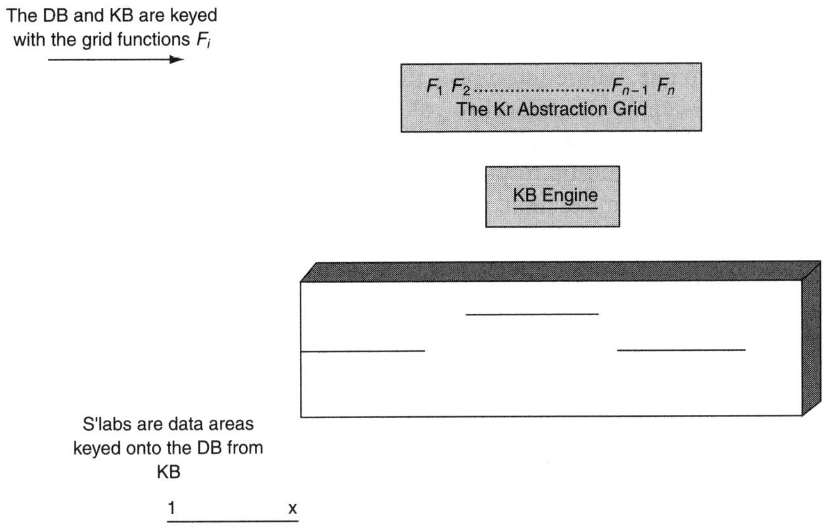

Figure 2. Keyed KR, inference, and visual discovery.

approach correspond to proof trees that have free Skolemized trees in their representation. While doing proofs with free Skolemized trees we are facing proofs of the form $p(g(\ldots))$ proves $p(f(g(\ldots)))$ and generalizations to $p(f(x))$ proves for all x, $p(f(x))$. Thus the free proofs are in some sense an abstract counterpart of the SLD. Let us see what predictive diagrams do for knowledge discovery knowledge management. Diagrams allow us to model-theoretically characterize incomplete KR. To key into the incomplete knowledge base. The following figure depicts selector functions F_i from an abstract view grid interfaced via an inference engine to a knowledge base and in turn onto a database.

Generalized predictive diagrams are defined, whereby specified diagram functions and search engine can select onto localized data fields. A generalized predictive diagram, is a predictive diagram where D(M) is defined from a minimal set of functions. The predictive diagram could be minimally represented by a set of functions $\{f_1, \ldots, f_n\}$ that inductively define the model. The functions are keyed onto the inference and knowledge base to select via the areas keyed to, designated as Si's in Figure 1 and data is retrieved Nourani (1999f). Visual object views to active databases might be designed with the above. The trees defined by the notion of provability implied by the definition might consist of some extra Skolem functions $\{g_1, \ldots, g_n\}$, that appear at free trees. The f terms and g terms, tree congruences, and predictive diagrams then characterize deduction with virtual trees Nourani (1999c) as intelligent predictive interfaces.

4.5. KR and Data Warehousing with Keyed Functions

Practical AI systems are designed by modelling AI with facts, rules, goals, strategies, knowledge bases. Patterns, schemas, AI frames and viewpoints are the micro to aggregate glimpses onto the database and knowledge bases were masses of data and their relationships-representations, respectively, are stored. Schemas and frames are what might be defined with objects, the object classes, the object class inheritances user-defined inheritance relations, and specific restrictions on the object, class, or frame slot types and behaviors. From Nourani (2000) scheme might be intelligent forecasting.

IS-A stock forecasting technique
Portfolios stock, bonds, corporate assets
Member management science techniques

Schemas allow brief descriptions on object surface properties with which high level inference and reasoning with incomplete knowledge can be carried out applying facts and the defined relationships amongst objects. Relationships: Visual Objects have mutual agent visual message correspondence. Looking for patterns is a way some practical AI is a carried on with to recognize important features, situations, and applicable rules. From the proofs standpoint patterns are analogies to features as being leaves on computing trees. Forward chaining is a goal satisfaction technique, where inference rules are activated by data patterns, to sequentially get to a goal by apply the inference rules. The current pertinent rules are available at an agenda store. The carried out rules modify the database. Backward chaining is an alternative based on opportunistic response to changing information. It starts with the goal and looks for available premises that might be satisfied to have gotten there. Goals are objects for which there is automatic goal generation of missing data at the goal by recursion backward chaining on the missing objects as sub-goals. Data unavailability implies search for new goal discovery. Goal directed planning is carried out while planning with diagrams. The part of the plan that involves free Skolemized trees is carried along with the proof tree for a plan goal. If the free proof tree is constructed then the plan has an initial model in which the goals are satisfied. IM's basis for forecasting is put forth at preliminary stages in Nourani (1999a). The Morph-Gentzen logic with predictive model diagrams has been applied as a basis for intelligent forecasting Nourani (1999a, 2000). The graphics sequents for predicting the quarter earnings from the second and third are combined with a market condition graph. The way a market condition graph is designed is a propriety issue. It is obtained by Morph Gentzen sequents from known stock market parameters. Data discovery from KR on diagrams might be viewed as satisfying a goal by getting at relevant data which instantiates a goal. The goal formula states what relevant data is sought.

4.6. Competitive Models and Multiplayer Games

The game trees are applied to train models. The computing model is based on a novel competitive learning with agent multiplayer game tree planning. Specific agents are assigned to transform the models to reach goal plans where goals are satisfied based on competitive game tree learning. The planning applications include DM, OR—Operations research as goal satisfiability and micro-managing decision support with means-end analysis.

Planning is based on goal satisfaction at models. Multiagent planning, for example as Muller-Pischel (1994), Brazier et al. (1997), in the paper is modeled as a competitive learning problem where the agents compete on game trees as candidates to satisfy goals hence realizing specific models where the plan goals are satisfied. When a specific agent group "wins" to satisfy a goal the group has presented a model to the specific goal, presumably consistent with an intended world model. For example, if there is a goal to put a spacecraft at a specific planet's orbit, there might be competing agents with alternate micro-plans to accomplish the goal. While the galaxy model is the same, the specific virtual worlds where a plan is carried out to accomplish a real goal at the galaxy via agents are not. Therefore, plan goal selections and objectives are facilitated with competitive agent learning. The intelligent languages (Nourani 1993) are ways to encode plans with agents and compare models on goal satisfaction to examine and predict via model diagrams why one plan is better than another or how it could fail. Virtual model planning is treated in the author's publications where plan comparison can be carried out at VR planning (Nourani 2001).

4.7. The Stock Traders Interface Model

The basis for forecasting is put forth at preliminary stages in the author's publications since 1998. The idea is to apply Morph-Gentzen logic as a basis for intelligent multimedia forecasting. The figure indicates a graphics sequent for predicting the fourth quarter earnings

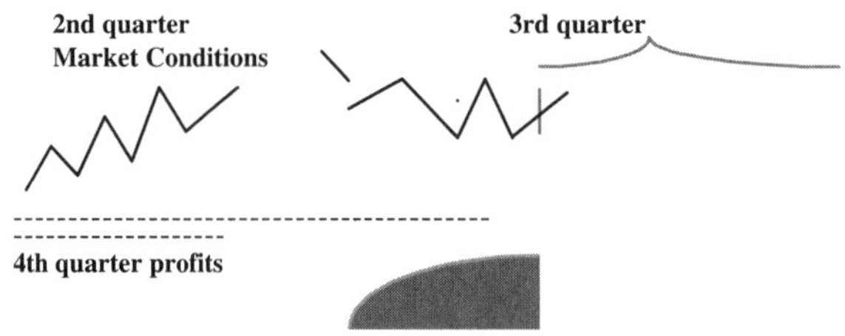

Figure 3. Stock forecasting.

from the second and third combined with a market condition graph. The way a market condition graph is designed is a propriety issue. It is obtained by Morph Gentzen sequents from known stock market parameters. The enclosed are example stock trading and forecasting interfaces.

4.8. The Visual Database Asset Management Model

The fast track to real media content management is a media asset management environment that allows vendors and customers media applications to interoperate seamlessly within the IBM DB2 digital library infrastructure. Asset management is an important area to the efficient intelligent multimedia computing. Cost-effective management of digital content is expected to be another frequent topic. IBM's grand central media "Grand Central Media" provides media asset management that enables a company to manage, secure, share, locate and reuse digital media files. The IBM solution also allows media and entertainment executives to import and index digital content and combine search requests into one results list that can be graphically browsed and shared with others world-wide. IBM grand central media is part of the company's suite of digital library applications that are being used by leading film studios, advertising firms, and broadcast and stock footage houses. The Web-based solution supports the synchronization of databases at different physical locations or on wide-area networks. The academia for example, the University of California, Berkeley's BMRC has designed a video on demand system with asset management.

4.9. Multimedia Middleware Databases UI Wrappers

Large-scale multimedia information systems today rely on large collections of data stored in diverse systems with differing capabilities. Many of the "mission-critical" tasks these customers do depend on integrating data from several sources. Often the data sources to be integrated include a broad range of sources, such as CAD/CAM systems, text search engines, molecular structure databases, or customer-specific repositories. Garlic (1995) offers the ability to interrelate data from multiple sources with a broad range of querying capabilities, in a single, cross-source query. Garlic is an IBM prototype, which allows new sources to be easily added to an existing installation. Garlic offers the ability to interrelate data from multiple sources with a broad range of querying capabilities, in a single, cross-source query. A significant focus of the project is the provision of support for data sources that provide type-specific indexing and query capabilities. Garlic's "wrapper architecture," see e.g., Wiederhold (1992) and the applications at Nourani (1996) encapsulate data sources, allowing new sources to be added quickly, and accommodating a wide variety of sources, with a broad range of traditional and nontraditional query processing capabilities. Wrappers model legacy data as objects, participate in query planning, and provide standard interfaces for method invocation and query execution. The knowledge base is keyed with relevant selector functions. There are worldwide web retrieval designs, which have to apply direct links. The hierarchical abstract structures are being applied to the knowledge base and data structures to accomplish the tasks called for.

5. THE APPLICATIONS TO DM AND KNOWLEDGE MANAGEMENT

5.1. Agent DM

Agent data mining starting with Figure 1, applies agents and intelligent multimedia, agent correlation signatures, processes, and virtual correlation. Agent correlation signature matches are applied with pixel-based queries. The search engine keys are agent-based functions or standard functions. To explore virtual correlation signatures let us start with canonical correlation analysis. The method is concerned with describing linear dependencies between sets of variables. Since we apply agent variables and virtual trees, we can apply projection operators to specific function parameterized trees to describe the relations between canonical vectors and variates. A kernel feature space is presented spanned by objects as in ordinary canonical correlation.

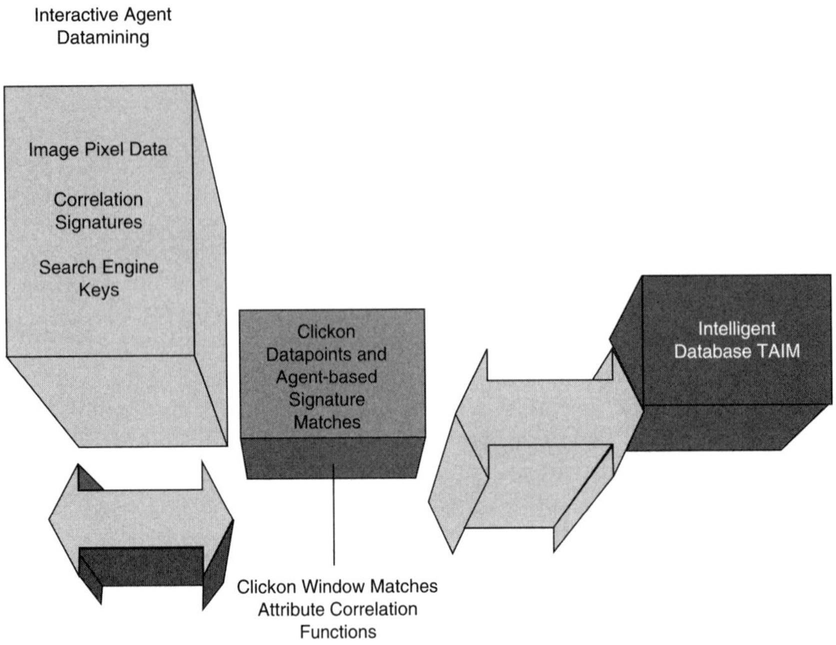

Figure 4. Agent datamining.
Note: The Design is the author's venture property. Applications must be with venture partnerships and permission.

The configuration is based on Figure 1, however, the queries are based on intelligent multimedia encodings, the search engine keys apply agent processes, and the correlation signature matches are on agent signatures.

5.2. TAIM-based Agent DataWarehouse

Agent data warehousing applies agent processes, and agent correlation signatures. Agent correlation signature matches are applied with pixel-based queries. Click on attribute data points and attribute correlation agent matches are applied. The search engine keys are

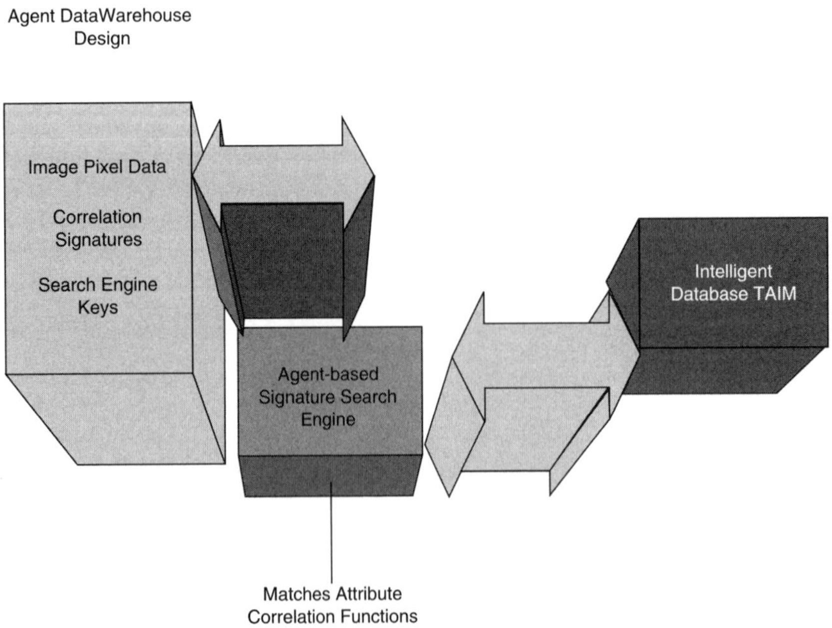

Figure 5. Agent datawarehousing.
Note: The Design is the author's venture property. Applications must be with venture partnerships and permission.

agent-based functions or standard functions. The correlation signature matches are on agent signatures.

5.3. Correlation Signatures

With agent computing and Morph Gentzen models correlation signatures can be computed and encoded with agents on objects. Thus there are agent correlation signatures. Agent signatures allow us to vary the content, objects, and pixel correlations. Canonical agent multimedia model computations are permeated on several chapters, in particular, in the mathematical foundations chapters and on a paper on a new geometric logic (Nourani 2002). Example application areas for the above is spatial searches, pixel-based queries, and retrieving the information stored in digital libraries containing either images or video, which is difficult to search or browse.

Studies of the correlations in images and their computations are of great interest and importance. Statistical analysis of natural images was pioneered shortly after development of information theory by Shannon.1 Initial studies were on television applications in which one-pixel probability distribution and two-pixel correlation functions, as well as two- and three-pixel probability distributions, 3 were obtained for television frames. The results were used to estimate the information content of the images. Further enhancements created with minimization of higher-order correlations. However, to make this claim one must demonstrate that the removal of higher-order dependencies results in a substantial increase in the efficient coding benefits.

The agent model allows us to encode images with functions on visual objects where the higher order dependency computations are minimized. The benefits are based on the reduction in information redundancy which, besides alleviating the bottleneck problem, decreases the number of possible associations as well as the amount of "vocabulary" necessary for pattern recognition. Further more as indicated in Chapter 5, the possible worlds are encoded with the image diagram functions. The redundancy due to the third- and higher-order correlations in natural images is large when measured in terms of coding bits, which is what matters for data compression and coding efficiency.

5.4. Knowledge Management (KM)

A basic understanding of the link between visual and textual information is necessary so that media processing technologies can be enhanced to extract and combine high level representations of media content. The representations must facilitate more intuitive and personalized access and navigation of media artifacts for users. To develop a computational account of how video and text can be integrated by representations of narrative in multimedia systems (Salway et al. 2003). Elaborate storytelling can occur in various combinations of media and it is interesting to consider how the same story can be conveyed and understood in, for example, a moving image and in spoken or written language. The above is beyond what ordinary story understanding systems can process. Consideration of the relationship between different kinds of media is also required for the generation of enriched multimedia information sources.

The need to integrate media representations, at different levels of abstraction, leads to the idea of a media interlingua (Maybury 1997). For example, low-level integration may involve representations of pixel distributions, sound wave features and keyword frequencies that can be expressed numerically. Higher level 'semantic media content' is generally equated with named events, objects, their attributes and their spatio-temporal relationships for which various logics have been proposed in the multimedia literature. However there are higher levels still that are not yet dealt with in multimedia systems. For example there are potential applications for which it may be necessary to represent sequences of connected events, causal relationships between events and the intentions of the people involved.

Whilst media such as films are an obvious example, this may also be relevant for other kinds of systems, for example tracking business news. We suggest that theories about narrative, alongside existing knowledge representation schemes, can guide the development of media independent representations to integrate media at this level. Textual information from

TV sit-com scripts was combined with visual features, through user interaction, in order for a system to locate scenes containing a particular character (Picard 1997) is produced specifically to be informative about the important objects, characters and events depicted by the moving image.

Many technologies can form components of an intelligent knowledge management system for unstructured data. These may include the following: *information retrieval*, attempting to satisfy a user's information need by locating documents containing relevant information; *summarization*, content reduction by specialization or generalization of a document; *information extraction*, locating facts from within a document and potentially inferring other information; *machine translation*, translation of content between one or more languages; *content indexing*, speech or image recognition and analysis to identify the content of multimedia data; and *information visualization*, to represent the contents of information sources or their interrelationships.

A number of other areas are of interest in knowledge management are areas where the content abstraction techniques in the preceding chapters can be applied language content. All these issues and many more, and the technical difficulty of achieving any of these indexing objectives, mean that there is currently little standardization in video indexing beyond detection of the boundaries between scene changes. Systems are typically developed to serve individual domains and often require extensive modification to be effective with new tasks.

5.4.1. Applications

There are many potential application areas for intelligent multimedia knowledge management technologies. The increasing availability of audio-visual recording equipment and the ease with which recordings can be made available online via the World Wide Web, means that it is becoming common practice to routinely record presentations such as lectures and seminars. One important application of this technology is the recording of university and college lectures. These are often made available online by lecturers for private study by those attending the lectures, but are also increasingly being utilised for distance learning by students physically remote to the institution. At present recordings are generally made available for playback in the manner of a convention video recording, for example features such as fast forward and preview are often supported. One important feature of recordings of this type is that they often contain material not otherwise available. Information delivered by speakers will often be unplanned, particularly when answering questions or taking part in discussions. These elements of a presentation are sometimes referred to as containing *tacit knowledge*; information which is based on personal experiences, know how, values and beliefs (Brown et al. 2001). Being able to automatically search these presentations will make these valuable resources much more accessible. The technology used to make these recordings varies greatly. Well funded lecture halls are often equipped with multiple high quality audio and video capture points and record all images projected onto screens. More often though a single portable video camcorder or audio recorder is used to capture the presentation.

5.4.2. Indexing Audio and Video Content

The hard question for which there is not a simple answer in the technologies outside the present book is, what should be indexed? Ideally all aspects of the image should be captured, but what are the essential aspects? Examples are to identify the type of location and perhaps uniquely identify it. We could attempt to recognize all the objects in the scene and go further to track their motion in the scene. How should an object to classified? Should a car be classified as a car or a vehicle, and need it include attributes such as colour, manufacturer, model, etc.? At present there seems to be little work in the development of standardized image label taxonomies. Of course, taxonomies of this type will be of little value if robust feature extraction is not available.

Automatic management of KM is one interesting feature in image analysis beyond classifying the natural scene is to recognize overlaid text. This material is often added to news broadcasts and other documentary material to identify locations or speakers. OCR techniques can be adapted to video images to identify this text which can then be used to help label names of speakers or locations. Education is a potentially important application area for intelligent multimedia knowledge management. One source of knowledge is archived recordings of audio presentations including lectures and seminars. Recordings of presentations often contain multiple information streams involving visual and audio data. If the full benefit of these recordings is to be realized these multiple media streams must be properly integrated to enable rapid navigation. This paper describes an approach to the integration of the audio soundtrack with electronic slides from a presentation by automatically synchronizing them. A novel component of the system is the detection of sections of the presentation unsupported by prepared slides, such as discussion and question answering, and automatic development of keypoint slides for these elements of the presentation (Grosky and Stanchev 2000).

6. CONCLUSIONS

Prespecified views are designed and obtained with intelligent multimedia active databases keyed with functions onto KR with generalized diagrams. As a science IM and Morph Gentzen are developing concepts and vocabulary to help us understand intelligent multimedia. The overview to a multimedia language, a logic—the Morph Gentzen logic and a brief view to its models are presented. The techniques are the basis for encoding intelligent multimedia systems and active databases with Morph channels. The knowledge base consists of behavior descriptions, vocabulary definitions, visual objects and relations, decision rules and uncertain facts. Generalized diagrams are shown to be an encoding for a minimal efficient KR technique applied to define relevant world models and KR for data search and discovery. Model checking is simplified to the same degree. New areas where the paper applies to is heterogeneous computing Nourani (1999) and mediation Wiederhold (1992). Proof abstraction is yet another area where KR can be keyed to KB and applied to knowledge discovery. We have defined various notions of generalized diagrams for KR applications and KB data discovery. KB completions can be applied with our techniques in Nourani (1999a, b). Multimedia datawarehouisng and discovery computation techniques with predictive diagrams are briefed. There are management science areas applied to on projects with colleagues at U Auckland being reported at conferences (Nourani and Loo 2000). User modeling is a prerequisite to usability. "Placing Trust in Human-Computer Interaction" (Arion et al. 1994) is a start for anyone who is trying to establish a usability presence. Topics include: making a business case for usability, overcoming inertia in large organizations, integrating usability into system development, and cultivating an effective client relationship. Multimedia data mining techniques and applications with agents were presented and specific correlation signature modeling. Areas on agent correction and virtual kernels are presented. Correlation encoding applications with functions and agents to search and discovery are stated. Agent warehousing models are further presented.

EXERCISES

1. Figure 2 Keyed KR, Inference, and Visual Discovery, how might the planning techniques in the preceding chapters be applied to search the database based on keyed functions applied to reach a plan goal on the KB?
2. Let $\mathbf{x}_i \in X$ $i = 1, \ldots, m$ denote input space objects and consider a feature space mapping $\phi\colon X \to F$ where the feature space F is an inner product space. The "kernel trick" is to calculate the inner product in F, $k(\mathbf{x}_i, \mathbf{x}_j) = \langle \phi(\mathbf{x}_i), \phi(\mathbf{x}_j) \rangle F$, using a kernel function $k\colon X \times X \to R$ of input space objects while avoiding explicit mappings ϕ.
   - 2a. *Apply agent techniques* to define an agent correlation kernel with x_i visual agent space objects.
   - 2b. Can you compute virtual correlations with varying objects on agents? What is an example virtual correlation?
3. In histogram refinement the pixels of a given bucket are subdivided into classes based on local features. There are many possible features, including texture, orientation, distance from the nearest edge, relative brightness, etc.

3a. Show that histogram refinement approximates the joint distribution of a variety of local properties.

3b. How do you prevent pixels in the same bucket from matching each other if they do not fall into the same class?

4. 4a. A number of other areas are of interest in knowledge management are areas where the context abstraction techniques in the preceding chapters can be applied.

4b. How are the context abstraction techniques applicable to *metadata* or *manual annotation* of the content?

4c. Can you state examples where intelligent language signatures and content abstraction can be applied to identify the document content or adding additional information beyond the content itself, identification of *context*, e.g. identifying temporal or location information associated with information, and *user interaction* with media, the platform to which it is being delivered, and user tasks.

5. An earlier study has measured three-pixel redundancy in terms of coding bits, and suggested that it gives a small contribution to the overall redundancy. However, this study ignored the two-dimensional nature of the images and accounted only for the redundancy arising from three neighboring pixels aligned horizontally. Furthermore, it performed the measurement for an ensemble of only a few small television frames at one level of image intensity quantization (64 shades of gray).

5a. Show that information redundancy depends strongly on the gray-level quantization, or equivalently on the number of intensity bits per pixel (pixel depth).

5b. There are mathematical framework accounting for a full set of interpixel redundancies in a two-dimensional array of pixels (c.f. Petrove et al. 2002). What are example models.

5c. Can you apply visual diagram function encoding to the above to minimize computations on correlations.

6. Applying the model in Section 4 to process high-level database queries.

6a. How do you encode a query with the functions F_i?

6b. How do you apply the KB as an inference engine to localize to specific items?

6c. Apply 6b to search in a digital arts gallery, for example, pertaining to certain images.

REFERENCES

1. M. Arion, J. H. Numan, H. Pitariu, and R. Jorna, "Placing Trust in Human-Computer Interaction." Proc. of the 7th European Cognitive Ergonomics Conference, 1994.
2. J. Bailey, M. Georgeff, D. B. Kemp, D. Kinny, and K. Ramamohuarao, "Active Databases and Agent Systems A Comparison," Technical Report: 95/10. University of Melbourne, Parkville, 1995.
3. C. Böhm, B. Braunmüller, H-P. Kriegel, and M. Schubert, "Efficient Similarity Search in Digital Libraries," pp. 193–206. Proc. of IEEE Advances in Digital Libraries (ADL'2000). Washington, DC, 2000.
4. G. Brajnik, G. Guida, and C. Tasso, *IEEE Transaction on Systems, Man, and Cybernetics* 20, 166 (1990).
5. D. Browne, "STUDIO: Structured User-interface Design for Interaction Optimization." Prentice Hall, New York, NY, 1994.
6. K. Chakrabarti, M. Ortega-Binderberger, K. Porkaew, P. Zuo, and S. Mehrotra, "Similar Shape Retrieval in MARS," pp. 709–712. IEEE Int. Conf. on Multimedia and Expo (ICME II), New York, NY, USA, 2000.
7. K. K. T. Cheung, R. W. K. Lam, H. H. S. Ip, L. H. Y. Tang, and R. Hanka, "A Software Framework for Combining Iconic and Semantic Content for Retrieval of Histological Images," pp. 488–499. In 4th Int. Conf. Visual, Lyon, France, 2000.
8. V. Chitkara, M. A. Nascimento, and C. Mastaller, "Content-Based Image Retrieval Using Binary Signatures," Technical report TR00-18, University of Alberta Edmonton, Alberta, Canada, 2000; S. Lin. "An Extendible Hashing Structure for Image Similarity Searches," Technical report 00-04, University of Alberta, Canada, 2000.
9. W. F. Cody, L. M. Haas, W. Niblack, M. Arya, M. J. Carey, R. Fagin, D. Lee, D. Petkovic, P. M. Schwarz, J. Thomas, M. Roth, J. H. Williams, and E. L. Wimmers, "Querying Multimedia Data from Multiple Repositories by Content: The Garlic Project," Proceedings IFIP 2.6 Third Working Conference on Visual Database Systems (VDB-3). Chapman & Hall, Lausanne, Switzerland, 1995.
10. K. Dekker and V. Lesser, "Analyzing the Need for Metalevel Communications," in Proc. of the Twelfth International Workshop on Distributed AI, DAI 93. Hidden Valley, Pennsylvania, 1993.
11. E. M. Del Galdo and J. Nielsen, Eds., "International User Interfaces." Wiley, New York, NY, 1996.
12. C. Djeraba, "When Image Indexing Meets Knowledge Discovery," Workshop on Multimedia Data Mining (MDM/KDD2000)—6th ACM SIGKDD Int. Conf. on Knowledge Discovery & Data Mining, Boston, MA, USA, 2000.

13. E. A. El-Kwae and M. R. Kabuka, *ACM Transactions on Information Systems* 18, 171 (2000).
14. M. Genesereth and N. J. Nilsson, "Logical Foundations of Artificial Intelligence." Morgan-Kaufmann, 1987.
15. G. Gentzen, *Math. Ann.* 119, 140 (1943).
16. V. Gouet, P. Montesinos, R. Deriche, and D. Pelé, "Evaluation de détecteurs de points d'intérêt pour la couleur," Reconnaissance des Formes et Intelligence Artificielle (RFIA '2000). Paris, 2000.
17. W. I. Grosky and P. Stanchev, "An Image Data Model" (R. Laurini, Ed.), Vol. 1929, pp. 14–25. Advances in Visual Information Systems, Lecture Notes in Computer Science, Springer-Verlag, 2000.
18. D. J. Guan, C-Y. Chou, and C-W. Chen, *Information Processing Letters* 75, 113 (2000).
19. K. Hirata, S. Mukherjea, W-S. Li, and Y. Hara, *Multimedia Tools and Applications* 11, 295 (2000).
20. R. Honda, H. Takimito, and O. Konoshi, "Semantic Indexing and Temporal Rule Discovery for Time-series Sattelite Images," Proc. for MultiMedia Data Mining (MDM/KDD 2000) Workshop—6th ACM SIGKDD Int. Conf. on Knowledge Discovery & Data Mining. Boston, MA, USA, 2000.
21. Wynne Hsu, T. S. Chua, and H. K. Pung, "An Integrated Color-spatial Approach to Content-based Image Retrieval," pp. 305–313. ACM Multimedia Conference, 1995.
22. A. Jaimes and S-F. Chang, "A Conceptual Framework for Indexing Visual Information at Multiple Levels in Internet Imaging," IS&T/SPIE, 2000a.
23. A. Jaimes and S-F. Chang, "Automatic Selection of Visual Features and Classifiers, in Storage and Retrieval for Media Databases," IS&T/SPIE, 2000b.
24. A. Kitamoto, "The Development of Typhoon Image Database with Content-Based Search," pp. 163–170. Proc. of the 1st Int. Symp. on Advanced Informatics (AdInfo), 2000.
25. Malte Kuss and Thore Graepel, "The Geometry of Kernel Canonical Correlation Analysis," Max–Planck-Institut für biologische Kybernetik, Technical Report No. 108, Max Planck Institute for Biological Cybernetics, Germany, 2003.
26. C. Largouet, "Aide à l'interprétation d'une séquence d'images par la modélisation du systè observé. Application à la reconnaissance de l'occupation du sol," Université de Rennes 1, France, 2000.
27. D. W. Loveland, "Automated Theorem Proving: A Logical Basis." North Holland, 1978.
28. M. T. Maybury, "Intelligent Multimedia Interfaces." MIT Press, 1997.
29. S. Mika, G. Rätsch, J. Weston, B. Schölkopf, and K.-R. Müller, "Fisher Discriminant Analysis with Kernels" (Y.-H. Hu, J. Larsen, E. Wilson, and S. Douglas, Eds.), pp. 41–48. Neural Networks for Signal Processing IX, IEEE, 1999.
30. Muller-Pischel, "ECAI 11th European Conference Artificial Intelligence" (A Cohn, Ed.), pp. 709–713. John Wiley and Sons, 1994.
31. C. F. Nourani, "Planning and Plausible Reasoning in AI," pp. 150–157. Proc. Scandinavian Conference in AI, IOS Press, Roskilde, Denmark, 1991.
32. C. F. Nourani, "Double Vision Computing." Proc. IAS-4, Intelligent Autonomous Systems. Karlsruhe University, Karlsruhe, Germany, 1993.
33. C. F. Nourani, "The IOOP Project," Vol. 30, pp. 54–56. SIGPLAN Notices, 1995a.
34. C. F. Nourani, "Free Proof Trees and Model-Theoretic Planning." AISB, Sheffiled, 1995b.
35. C. F. Nourani, "AII and Heterogenous Software Design," Eighth European Workshop on Modelling Autonomous Agents in a Multi-Agent World, University of Karlskrona/Ronneby, Ronneby, Sweden, 1996.
36. C. F. Nourani, "Agent Computing, KB for Intelligent Forecasting, and Model Discovery for Knowledge Management," AAAI Wkshp on Agent Based Systems in the Business Context. Orlando, Florida, 1999a.
37. C. F. Nourani, "KR and Model Discovery with Predictive Logic," Australian AI Conferene, AKA Wkshop, Sydney, 1999b.
38. C. F. Nourani, "Virtual Tree Computing" Abstract Posted at the ECCAD, Carolinas, 1999c.
39. C. F. Nourani, "Intelligent Multimedia—New Computing Techniques and Its Applications" (Ch. Freytag and V. Wolfengagen, Eds.). CSIT'99, Proceedings of 1st International Workshop on Computer Science and Information Technologies. MEPhI Publishing, Moscow, Russia, 1999d.
40. C. F. Nourani, "Intelligent Multimedia—Computing Techniques and Design Paradigms." Treeless Press, Berkeley, California, 1999e.
41. C. F. Nourani, "AII, Heterogenous Design, and Retrieval Agents" Brief Version at IJCAI-99, IIIS Track, Stokholm, 1999f.
42. C. F. Nourani, "Intelligent Heterogenous Multimedia and Intelligent Active Visual DB and KB," ICCIT, Las Vegas, Nevada, 2000a.
43. C. F. Nourani, "Morph Gentzen, KR, and World Model Diagrams," Automated Deductions and Geometry. Zurich, 2000b.
44. C. F. Nourani and G. S. L. Loo, "Intelligent Business Objects and Agent Computing." Parallel and Distributed Computing, University of Auckland, Las Vegas, Nevada, 1999.
45. C. F. Nourani and G. S. L. Loo, "KR and Model Discovery from Active DB with Predictive Logic," Data Mining 2000, Applications to Business and Finance. The University of Auckland, Cambridge, UK, 2000.
46. C. F. Nourani and R. M., "Morph Gentzen Plan Computation with Visual Diagrams," ISTA the Conference Proceedings published in the GI-Edition Lecture Notes in Informatics. Kharkiv, Ukraine, 2001.
47. C. F. Nourani, "Morph Gentzen, KR, and World Model Diagrams" Automated Deductions and Geometry, Zurich, 2002.
48. Virginia Ogle and Michael Stonebraker, *IEEE Computer* 28, 40 (1995).
49. JH. Oh and K. A. Hua, "Efficient and Cost-effective Techniques for Browsing and Indexing Large Video Databases," pp. 415–426. Proc. of ACM SIGMOD, Int. Conf. on Data Management. Dallas, Texas, 2000.

50. V. Oria, M. Tamer Özsu, P. J. Iglinski, S. Lin, and B. Yao, "DISIMA: A Distributed and Interoperable Image Database System," p. 600. Proc. of ACM SIGMOD, Int. Conf. on Data Management. Dallas, Texas, 2000.
51. Alex Pentland, Rosalind Picard, and Stan Sclaroff, *International J. Computer Vision* 18, 233 (1996).
52. Petra Perner "Data Mining on Multimedia Data." Springer Verlag, Berlin-Heidelberg, 2002.
53. Yury Petrov and L. Zhaoping, "Local Correlations, Information Redundancy, and Sufficient Pixel Depth in Natural Images," Oxford University, Oxford, UK; University College London, London, 2002.
54. R. W. Picard, "Affective Computing." MIT Press, 1997.
55. D. Poole, R. Goebel, and R. Aleliunas "Theoreist: A Logical Reasoning System for Defaults and Diagnosis" (N. Cercone and G. McCalla, Eds.), The Knowledge Frontier: Essays in the Representation of Knowledge. Springer-Verlag, 1987.
56. D. Prawitz, "Natural Deduction: A Proof Theoretic Study." Almqvist and Wiksell, Stokholm, 1965.
57. K. Rodden, W. Basalaj, D. Sinclair, and K. Wood, "A Comparison of Measures for Visualising Image Similarity." Third UK Conference on Image Retrieval (CIR '2000), Brighton, UK, 2000.
58. R. Rosipal and L. J. Trejo, *J. Machine Learning Research* 2, 97 (2001).
59. A. Salway, M. Graham, E. Tomadaki, and Y. Xu, "Linking Video and Text via Representations of Narrative," p. 24. AAAI Symposium on Intelligent Multimedia Knowledge Management, Stanford. Palo Alto, 2003.
60. B. Schölkopf, A. Smola, and K.-R. Müller, *Neural Computation* 10, 1299 (1998).
61. B. Schölkopf and A. J. Smola, "Learning with Kernels," The MIT Press, Cambridge, Massachusetts, 2002.
62. S. J. Simoff, "Variations on Multimedia Data Mining." Proc. for Multimedia DataMining MDM/KDD2000 Workshop, in Conjunction with ACM SIGKDD Int. Conf. on Knowledge Discovery & Data Mining. Boston, MA, USA, 2000.
63. A. W. M. Smeulders, M. Worring, S. Santini, A. Gupta, and R. Jain, *IEEE Trans. on Pattern Analysis and Machine Intelligence* 22, 1349 (2000).
64. Steele and Pater Yanker, *IEEE Computer* 28, 23 (1995).
65. L. H. Y. Tang, R. Hankaa, H. H. S. Ipb, K. K. T. Cheungb, and R. Lamb, "Semantic Query Processing and Annotation Generation for Content-based Retrieval of Histological Images," Vol. 3980. Proc. of SPIE Medical Imaging: PACS Design and Evaluation: Engineering and Clinical Issues, 2000.
66. Sebastian Thrun, C. Faloutsos, T. Mitchell, and L. Wasserman, "Automated Learning and Discovery: State-of-The-Art and Research Topics in a Rapidly Growing Field." CMU-CALD-98-100. Center for Automated Learning and Discovery, Carnegie Mellon University, Pittsburgh, PA, 1998.
67. L. Trenner and J. Bawa, Eds., "The Politics of Usability: A Practical Guide to Designing Usable Systems in Industry." Springer-Verlag, London, UK, 1998.
68. T. van Gestel, J. A. K. Suykens, J. D. Brabanter, B. D. Moor, and J. Vandewalle, "Kernel Canonical Correlation Analysis and Least Squares Support Vector Machines" (G. Dorffner, H. Bischof, and K. Hornik, Eds.), pp. 381–386. Proceedings of the International Conference on Artificial Neural Networks, Springer, Berlin, 2001.
69. R. C. Veltkamp and M. Tanase, "Content-Based Image Retrieval Systems: A Survey," Technical report UU-CS-2000-34, Utrecht University, 2000.
70. Alexei Vinokourov, David R. Hardoon, and John Shawe-Taylor, "Learning the Semantics of Multimedia Content with Application to Web Image Retrieval and Classification." University of London, Royal Holloway, Egham, UK, 2004.
71. Gio Wiederhold, "Views, Objects, and Databases," Vol. 19, pp. 37–44. IEEE Computer, 1986.

CHAPTER 8

The Applications to Business and E-Commerce

CONTENTS

1. INTRODUCTION

Modeling, objectives, and planning issues are examined with agent planning and competitive models. Model discovery and prediction is applied to compare models and get specific confidence intervals to supply to goal formulas. Competitive model learning is presented, starting with the new agent computing bases defined since 1994. The foundations are applied to present precise decision strategies on multiplayer games with only perfect information between agent pairs. The game tree model is applied to train models. The computing model is based on a novel competitive learning with agent multiplayer game tree planning. Specific agents are assigned to transform the models to reach goal plans where goals are satisfied based on competitive game tree learning. Intelligent and/or trees and means-end analysis is applied with agents as the hidden–step computations.

A novel multiplayer game model is presented where “intelligent” agent enriched languages can be applied to address game questions on models in the mathematical logic sense. The new MIS as an academic and business field might be depicted by the enclosed figure. Software agents are specific agents designed by a language that carry out specified tasks and define software functionality. Most agents defined by our examples are software agents. Academic MIS essentials might be redefined as the figure indicates. There is agent computing, cyberspace computing, intelligent multimedia and heterogeneous computing. Plans

ISBN: 1-58883-037-3/$35.00

Intelligent Multimedia Computing Science
By Cyrus F. Nourani
Pages: 103–110

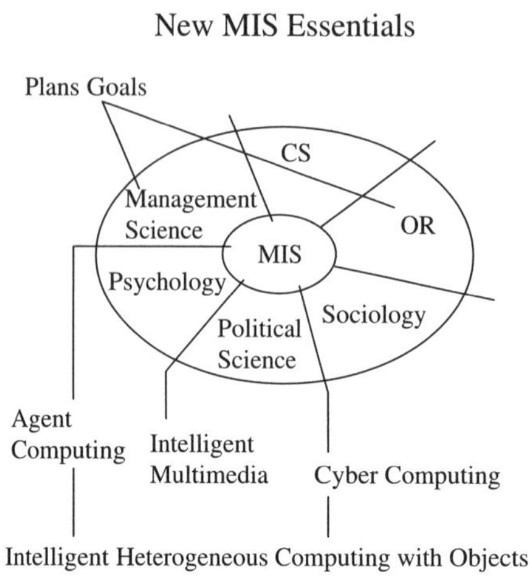

Figure 1. Agent-based business and MIS model.

and goals are in Figure 1. A preliminary multimedia forecasting plan is put forth in the author's papers. The idea is to apply Morph-Gentzen logic as a basis for intelligent multimedia forecasting Nourani (1999a). A graphics sequent is applied to predict the trends. Specific market condition graphs are obtained by Morph Gentzen sequents from known stock market parameters.

2. PLANNING WITH PREDICTION

Modeling with agent planning is applied where uncertainty, including effector and parameter uncertainty, are relegated to agents, where competitive learning on game trees determines a confidence interval. The incomplete knowledge modelling is treated with KR on predictive model diagrams. Model discovery at KB's are with specific techniques defined for trees. Model diagrams allow us to model-theoretically characterize incomplete KR. To key into the incomplete knowledge base we apply generalized predictive diagrams whereby specified diagram functions a search engine can select onto localized data fields. The predictive model diagrams (Nourani 1995) could be minimally represented by the set of functions $\{f_1, \ldots, f_n\}$ that inductively define the model. Data discovery from KR on diagrams might be viewed as satisfying a goal by getting at relevant data which instantiates a goal. The goal formula states what relevant data is sought. We propose methods that can be applied to planning with diagrams to implement discovery planning. In planning with G-diagrams that part of the plan that involves free Skolemized trees is carried along with the proof tree for a plan goal. computing with diagram functions allows us to key to active visual databases with agents. Diagrams are well-known concepts in mathematical logic and model theory. The diagram of a structure is the set of atomic and negated atomic sentences that are true in that structure.

Models uphold to a deductive closure of the axioms modelled and some rules of inference. The generalized diagram (G-diagram) Nourani (19991, 1999a) is a diagram in which the elements of the structure are all represented by a specified minimal set of function symbols and constants, such that it is sufficient to define the truth of world formulas only for the terms generated by the minimal set of functions and constant symbols. Such assignment implicitly defines the diagram. It allows us to define a canonical model in terms of a minimal family of function symbols. The minimal set of functions that define a G-diagram are those with which a standard model could be defined. Formal definition of diagrams are stated here, generalized to G-diagrams, and applied in the sections to follow.

2.1. Prediction and Discovery

Minimal prediction is an artificial intelligence technique defined since the author's model-theoretic planning project. It is a cumulative approximation attained with completing model diagrams on what might be true in a model or knowledge base. A *predictive diagram* for a theory T is a diagram D(M), where M is a model for T, and for any formula q in M, either the function $f: q \rightarrow \{0, 1\}$ is defined, or there exists a formula p in D(M), such that $T \cup \{p\}$ proves q; or that T proves q by minimal prediction. A *generalized predictive diagram*, is a predictive diagram with D(M) defined from a minimal set of functions. The predictive diagram could be minimally represented by a set of functions $\{f_1, \ldots, f_n\}$ that inductively define the model. The free trees we had defined by the notion of provability implied by the definition, could consist of some extra Skolem functions $\{g_1, \ldots, g_l\}$ that appear at free trees. The f terms and g terms, tree congruences, and predictive diagrams then characterize partial deduction with free trees. The predictive diagrams are applied to discover models to the intelligent game trees. Prediction is applied to plan goal satisfiability and can be combined with plausibility (Nourani 1991), probabilities, and fuzzy logic to obtain, for example, confidence intervals.

2.2. KR with Keyed Functions

Practical AI systems are designed by modelling AI with facts, rules, goals, strategies, knowledge bases. Patterns, schemas, AI Frames and viewpoints are the micro to aggregate glimpses onto the database and knowledge bases were masses of data and their relationships-representations, respectively, are stored. Schemas and frames are what might be defined with objects, the object classes, the object class inheritances, user-defined inheritance relations, and specific restrictions on the object, class, or frame slot types and behaviors.

A scheme might be as follows:

Intelligent Forecasting
IS-A Stock Forecasting Technique
Portfolios Stock, bonds, corporate assets
Member Management Science Techniques

Schemas allow brief descriptions on object surface properties with which high level inference and reasoning with incomplete knowledge can be carried out, applying facts and the defined relationships amongst objects. Relationships: Visual Objects A and B have mutual agent visual message correspondence. Looking for patterns is a way some practical AI is a carried on with to recognize important features, situations, and applicable rules. From the proofs standpoint patterns are analogies to features as being leaves on computing trees. *Forward chaining* is a goal satisfaction technique, where inference rules are activated by data patterns, to sequentially get to a goal by apply the inference rules. The current pertinent rules are available at an *agenda store*. The carried out rules modify the database.

Backward chaining is an alternative based on opportunistic response to changing information. It starts with the goal and looks for available premises that might be satisfied to have gotten there. Goals are objects for which there is automatic goal generation of missing data at the goal by recursion backward chaining on the missing objects as sub-goals. Data unavailability implies search for new goal discovery. Goal Directed Planning is carried out while planning with diagrams. That part of the plan that involving free Skolemized trees is carried along with the proof decision tree for a plan goal. If the free proof tree is constructed then the plan has a standard computable model in which the goals are satisfied. Let us see what predictive diagrams do for knowledge discovery knowledge management. Diagrams allow us to model-theoretically characterize incomplete KR. To key into the incomplete knowledge base. Selector functions F_i from an abstract view grid interfaced via an inference engine to a knowledge base and in turn onto a database. Generalized predictive diagrams whereby specified diagram functions a search engine can select onto localized data fields.

The functions are keyed onto the inference and knowledge base to select via the areas keyed onto.

Visual object views to active databases might be designed with the above. The trees defined by the notion of provability implied by the definition might consist of some extra Skolem functions $\{g_1, \ldots, g_n\}$, that appear at free trees. The f terms and g terms, tree congruences, and predictive diagrams characterize deduction with virtual trees as intelligent predictive interfaces. Data discovery from KR on diagrams might be viewed as satisfying a goal by getting at relevant data which instantiates a goal. The goal formula states what relevant data is sought. We have presented planning techniques, which can be applied to implement discovery planning. In planning with G-diagrams that part of the plan that involves free Skolemized trees is carried along with the proof tree for a plan goal. The idea is that if the free proof tree is constructed then the plan has a model in which the goals are satisfied. The model is the standard model of the AI world for which the free Skolemized trees were constructed.

3. COMPETITIVE MODELS AND GAMES

Planning is based on goal satisfaction at models. Multiagent planning, for example as Muller-Pischel (1994) and Brazier et al. (1997), here is modeled as a competitive learning problem where the agents compete on game trees as candidates to satisfy goals hence realizing specific models where the plan goals are satisfied. When a specific agent group "wins" to satisfy a goal the group has presented a model to the specific goal, presumably consistent with an intended world model. For example, if there is a goal to put a spacecraft at a specific planet's orbit, there might be competing agents with alternate micro-plans to accomplish the goal. While the galaxy model is the same, the specific virtual worlds where a plan is carried out to accomplish a real goal at the galaxy via agents are not. Therefore, plan goal selections and objectives are facilitated with competitive agent learning. The intelligent languages are ways to encode plans with agents and compare models on goal satisfaction to examine and predict via model diagrams why one plan is better than another or how it could fail. Virtual model planning is treated in the author's publications where plan comparison can be carried out at VR planning (Nourani 1997). Games play an important role as a basis to economic theories. Here the import is brought forth onto decision tree planning with agents.

Intelligent tree computing theories we have defined since 1994 can be applied to present precise strategies and prove theorems on multiplayer games. Game tree degree with respect to models is defined and applied to prove soundness and completeness. The game is viewed as a multiplayer game with only perfect information between agent pairs. Upper bounds on determined games are presented. The author had presented a chess-playing basis (Nourani 1997) to a computing conference. For each chess piece a designating agent is defined. The player P makes its moves based on the board B it views. ⟨P, B⟩ might view chess as if the pieces on the board had come alive and were autonomous agents carrying out two-person games as in Alice in Wonderland. Game moves are individual tree operations.

3.1. Intelligent AND/OR Trees and Search

AND/OR trees (Nilsson 1969) are game trees defined to solve a game from a player's stand point.

Formally a node problem is said to be solved if one of the following conditions hold.

1. The node is the set of terminal nodes (primitive problem—the node has no successor).
2. The node has AND nodes as successors and the successors are solved.
3. The node has OR nodes as successors and any one of the successors is solved.

A solution to the original problem is given by the subgraph of AND/OR graph sufficient to show that the node is solved. A program which can play a theoretically perfect game would have task like searching and AND/OR tree for a solution to a one-person problem to a two-person game. An intelligent AND/OR tree is and AND/OR tree where the tree branches are intelligent trees. The branches compute a Boolean function via agents. The Boolean function is what might satisfy a goal formula on the tree. An intelligent AND/OR tree is solved iff the corresponding Boolean functions solve the AND/OR trees named by intelligent functions on the trees. Thus node m might be $f(a_1, a_2, a_3)$ & $g(b_1, b_2)$, where f and g are Boolean functions of three and two variables, respectively, and a_i's and b_i's are Boolean valued agents satisfying goal formulas for f and g.

The chess game trees can be defined by agent augmenting AND/OR trees (Nilsson 1987). For the intelligent game trees and the problem solving techniques defined, the same model can be applied to the game trees in the sense of two person games and to the state space from the single agent view. The two person game tree is obtained from the intelligent tree model, as is the state space tree for agents. To obtain the two-person game tree the cross-board-coboard agent computation is depicted on a tree. Whereas the state-space trees for each agent is determined by the computation sequence on its side of the board-coboard. Thus a tree node m might be $f(a_1, a_2, a_3)$ & $g(b_1, b_2)$, where f and g are Boolean functions of three and two variables, respectively, and a_i's and b_i's are Boolean valued agents satisfying goal formulas for f and g.

A tree game degree is the game state a tree is at with respect to a model truth assignment, e.g. to the parameters to the Boolean functions above. Let generic diagram or G-diagrams be diagrams definable by specific functions. Intelligent signatures (Nourani 1996a) are signatures with designated multiplayer game tree function symbols. A soundness and completeness theorem is proved on the intelligent signature language (Nourani 1999a). The techniques allowed us to present a novel model-theoretic basis to game trees, and generally to the new intelligent game trees. The following specifics are from Nourani (1999c). Let N be the set of all functions from ω to ω. Let A be a subset of N. Gale-Stewart (1953) associated with A a 2-person game of perfect information G$\langle$A$\rangle$. Player I begins by choosing n_0 in ω; player two chooses n_1 in ω; then I chooses n_2 in ω; so on. Let $a_{(i)} = n_i$. I wins G$\langle$A$\rangle$ if and only if a in A. We say that G$\langle$A$\rangle$ is determined if one of the players has a winning strategy. To get a glimpse onto the specifics let us start with the following basics.

PROPOSITION. *If $G\langle A\rangle$ is determined, the complexity upper bound on the number of moves to win is A's cardinality.*

THEOREM. *For every pair p of opposing agents there is a set $A\langle p\rangle \subseteq N$. The worse case bound for the number of moves for a determined game based on the intelligent game tree model is the sum* $(\{|A\langle p\rangle|{:}p \text{ agent pairs}\}$.

PROOF. Sum over the proposition (Nourani 1999c).

At the intelligent game trees the winning agents determine the specific model where the plan goals are satisfied.

3.2. Two-Person Games

From the game tree view point for what Shannon had estimated a complete tree carried to depth 6-three moves for each player would already have one billion tip nodes. Yet from an abstract mathematical viewpoint only, the game is a two-person game with perfect information (Gale-Stewart 1953). However, the two-person game view is not a mathematical model for any chess playing algorithm or machine. The real chess game, from the abstract viewpoint, might well be modeled as a multiagent game, being only a two-agent game with perfect information between mutually informable agents.

A multiagent chess design or chess computing by any technique does not in reality provide perfect information in a way, which can be applied. The perfect information overall is a massive amount of data to be examined. There are thousands of move trees computed for searches coming close to being exhaustive. The multiagent chess paradigm (Nourani 1997) is not based on an exhaustive two-person game with perfect information model. There is only minimal information for the multiagent plans across the board. The multiagent multiboard model is a realization where the game is partitioned and correlated amongst agents and boards, with a cognitive anthropomorphism to human player's mind. There is an abstract two-person game model, but it does not apply to define a chess-playing machine. It is there to make precise mathematical statements.

3.3. Management Process Models

There are specific application areas for multiagent computing to Multinational corporation and their strategic management of multinational enterprises in Nourani (1998a). The areas to be applied to are global planning, external enterprise assessment, goal setting applications for operations research and market forecasting (Nourani 1991). The figures indicate the specific models starting with a multi-business corporate model.

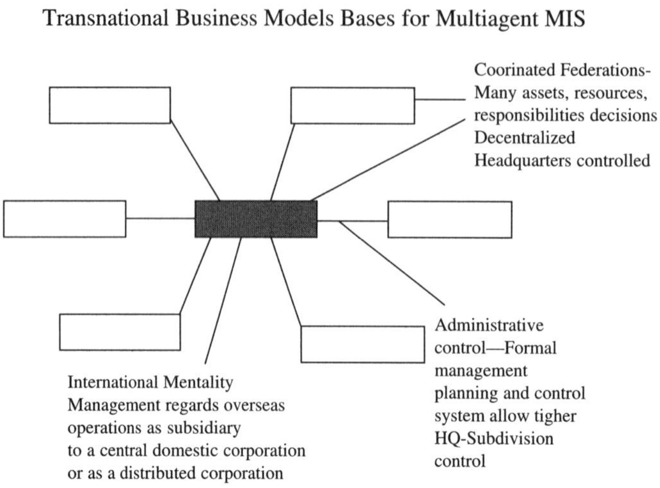

Figure 2. A Transactional agent MIS model.

The figures indicate the specific models starting with a multinational business model.

The organizational knowledge (van Heijst et al. 1996) is one of the main bases to competitive advantage. Enterprise modelling includes stock management, payroll, and advanced administrative tasks applying decision support. The following figure is a glimpse onto applying means-end analysis decision support where the hidden steps are designed and computed with parameter agents.

3.4. Decision Trees

We have defined specific application areas for multiagent computing to multinational corporations and their strategic management of multinational transactional business models appears in brief at Nourani (1998a). The areas applied to are global planning, external enterprise assessment, and goal-setting applications for operations research and market forecasting (Nourani 1998b). A specific models starting with a transactional business model is in Nourani (1997). The organizational knowledge (van Heijst et al. 1996) is one of the main bases to competitive advantage. Enterprise modelling includes stock management, payroll, and advanced administrative tasks applying decision support. The following figure is a glimpse onto applying means-end analysis decision support where the hidden steps are designed and computed with parameter agents. The obvious planning goal satisfaction applications are where agents apply backward chaining from objectives.

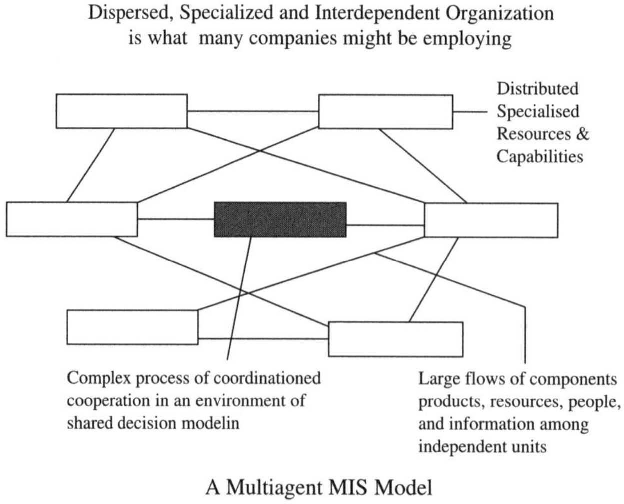

Figure 3. Multiagent MIS.

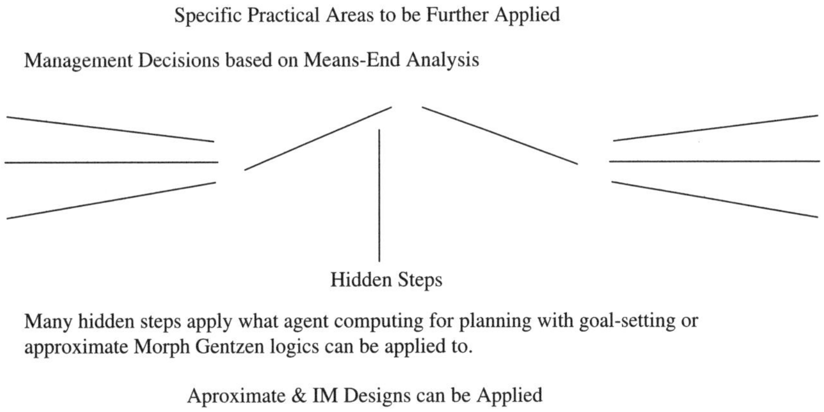

Figure 4. Means-end agent decision tree.

4. CONCLUSIONS

A novel basis to decision-theoretic planning is presented classical and non-classical planning techniques, see for example, Genesereth and Nilsson (1987), Wilkins (1984) from artificial intelligence with games and decision trees providing a agent expressive planning model. We use a broad definition of decision-theoretic planning that includes planning techniques that deal with all types of uncertainty and plan evaluation. Planning with predictive model diagrams represented with keyed KR to knowledge bases is presented. Techniques for representing uncertainty, plan generation, plan evaluation, plan improvement, and are accommodate with agents, predictive diagrams, and competitive model learning. Modeling with effector and sensor uncertainty, incomplete knowledge of the current state, and how the world operates is treated with agents and competitive models. Bounds on game trees are presented as a measure on the complexity of model comparison and competitive learning. Applications to means-end analysis and decision support with goal planning is presented.

EXERCISES

1. Model: A basis for forecasting is put forth at preliminary stages in the book, presented further here and Chapter 11. The idea is to apply Morph-Gentzen logic as a basis for intelligent multimedia forecasting. The figure indicates a graphics sequent for predicting the fourth quarter earnings from the second and third combined with a market condition graph. The way a market condition graph is designed is a propriety issue. It is obtained by Morph Gentzen sequents from known stock market parameters.

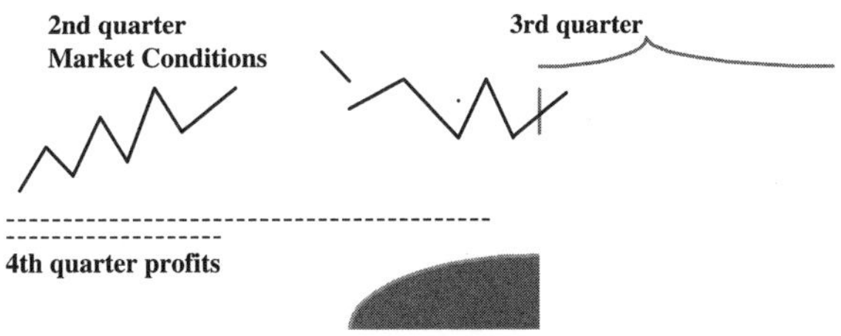

   Can you design some stock market condition graphs?
2. What are necessary market parameters?
3. What is an Intelligent Multimedia model for WWW stock market?
4. What business objects might be applied to define banking transactions?
5. Design a multimedia business portfolio.

REFERENCES

1. F. M. T. Brazier, B. Dunin-Keplicz, N. R. Jennings, and J. Treur, "DESIRE: Modelling Multi-Agent Systems in a Compositional Formal Framework" (M. Huhns and M. Singh, Eds.), Vol. 1. International Journal of Cooperative Information Systems, 1997.
2. D. Gale and F. M. Stewart, "Infinitite Games with Perfect Information," Vol. 28. In Contributions to the Theory of Games, Annals of Mathematical Studies, Princeton, 1953.
3. M. R. Genesereth and N. J. Nilsson, "Logical Foundations of Artificial Intelligence." Morgan-Kaufmann, 1987.
4. J. V. Hagel and A. G. Armstrong, "Net Gain—Expanding Market through Virtual Communities." Harvard Business School Press, Boston, Massachussetts, 1997.
5. K. Lauden and J. Lauden, *MIS Essentials*, Prentice-Hall, 1997.
6. J. P. Muller and M. Pischel, "Modeling Interactive Agents in Dynamic Environments," Proceedings 11th European Conference on AI, John Wiley and Sons, Amsterdam, The Netherlands, 1994.
7. C. F. Nourani, "Planning and Plausible Reasoning in AI," pp. 150–157. Proceedings Scandinavian Conference in AI, IOS Press, Denmark, 1991.
8. C. F. Nourani, "Intelligent Lanaguages—A Preliminary Syntactic Theory" (A. Kelemenová, Ed.), pp. 281–287. In Proceedings of the Mathematical Foundations CS'98 Satellite workshop on Grammar Systems. Silesian University, Opava, 1998.
9. C. F. Nourani, "Free Proof Trees and Model-Theoretic Planning." Automated Reasoning, AISB, England, 1995.
10. C. F. Nourani, "Slalom Tree Computing," Vol. 9, No. 4. AI Communications, 1996a.
11. C. F. Nourani, "Multiagent Chess Games," AAAI Chess Track, Providence, RI, 1997.
12. C. F. Nourani, "Agent Computing, KB for Intelligent Forecasting, and Model Discovery for Knowledge Management." AAAI—Workshop on Agent Based Systems in the Business Context, Orlando, Florida, 1999a.
13. C. F. Nourani, "Infinitary Multiplayer Games," Summer Logic Colloquium, Utrecht, 1999c.
14. N. J. Nilsson, "Searching, Problem Solving, and Game-playing Trees for Minimal Cost Solutions" (A. J. Morell Ed.), Vol. 2, pp. 1556–1562. IFIP 1968, North-Holland, Amsterdam, 1969.
15. N. J. Nilsson, "Problem Solving Methods in Artificial Intelligence." McGraw-Hill, New York, 1971.
16. G. van Heijst, R. Van der Speck, and E. Kruizinga, "Organizing Corporate Memories," In Proceedings of KAW, 1996.
17. D. Wilkins, *Artificial Intelligence* 22, 269 (1984).

CHAPTER 9

Spatial Heterogeneous Visual Computing

CONTENTS

1. INTRODUCTION

Multiagent visual computing techniques are applied to design autonomous flight control and spacecraft navigation. The basic technique introduced in Nourani (1983) is that of viewing the world as many possible worlds with agents at each world that compliment one another in problem solving by cooperating. The double vision computing paradigm with objects and agents might be depicted by the following figure. The cooperative problem solving paradigms have been applied ever since the AI methods put forth by Hays-Roth. However, the multiagent multiboard techniques due to the author (1983). The techniques to be presented are to

ISBN: 1-58883-037-3/$35.00

Intelligent Multimedia Computing Science
By Cyrus F. Nourani
Pages: 111–121

be applied to Mobile Multimedia. Communication and computation by multimedia visual-object languages can be programmed with IM with a simple syntax. The techniques to be presented are to be applied for (a) Precomputed video-object composition and combination for spatial Morph Gentzen computing with visual objects (b) High speed visual spacecraft navigation with multiagent multimedia.

The Morph Gentzen Computing Logic for computing for multimedia is new projects with important computing applications since Nourani (1998a). The basic principles are a mathematical logic where a Gentzen (1943) or natural deduction systems are defined by taking arbitrary structures and multimedia objects coded by diagram functions. Multimedia objects are viewed as syntactic objects defined by functions, to which the deductive system is applied. The deduction rules are a Gentzen system augmented by Morphing, and Transmorphing on events $\{p_1, p_2, \ldots, p_n\}$ with a consequent hybrid picture p. Thus the combination is an impetus event. The deductive theory is a Gentzen system in which hybrid pictures are named by parameterized functions; augmented with the Morph rules. A soundness and completeness theorem has been put forth in the author's papers on Morph Gentzen, which is applied to visual multiagent flight control and planning.

2. MULTIAGENT VISUAL PLANNING/LANNING

The visual field is represented by visual objects connected with agents carrying information amongst objects about the field, and carried onto intelligent trees for computation. Intelligent trees compute the spatial field information with the diagram functions. The trees defined have function names corresponding to computing agents. The computing agent functions have a specified module defining their functionality. Multiagent spatial vision techniques are introduced in the author's papers (1996). The multiagent multi-board techniques due to Nourani (1995a). The BID (Brazier-Treure et al. 1997) model has to be enhanced to be applicable to intelligent multimedia. Let us start with an example multi-board model where there multiagent computations based on many boards, where the boards corresponds to either virtual possible worlds or to alternate visual views to the world, or to the knowledge and active databases. The board notion is a generalization of the Blackboard problem-solving model since Hayes-Roth. The blackboard model consists of a global database called the blackboard and logically independent sources of knowledge called the knowledge sources. The knowledge sources respond opportunistically to the changes on the blackboard. Starting with a problem the blackboard model provides enough guidelines for sketching a solution. Agents can cooperate on a board with very specific engagement rules not to tangle the board or the agents. The multiagent multi-board model, henceforth abbreviates as MB, is a virtual platform to an intelligent multimedia BID agent-computing model. We are faced with designing a system consisting of the pair ⟨IM-BID, MB⟩, where IM-BID is a multiagent multimedia computing paradigm where the agents are based on the BID model. The agents with motivational attitudes model is based on some of the assumptions described as follows. Agents are assumed to have the extra property of rationality: they must be able to generate goals and act rationally to achieve them, namely planning, replanting, and plan execution. Moreover, an agent's activities are described using mentalists notions usually applied to humans. To start with the way the mentalists attitudes are modulated is not attained by the BID model. It takes the structural IM-BID to start it. The preceding chapters and sections on visual context and epidemics have brought forth the difficulties in tackling the area with a simple agent computing model. The BID model does not imply that computer systems are believed to actually "have" beliefs and intentions, but that these notions are believed to be useful in modeling and specifying the behavior required to build effective multi-agent systems. The first BID assumption is that motivational attitudes, such as beliefs, desires, intentions and commitments are defined as reflective statements about the agent itself and about the agent in relation to other agents and the world. These reflective statements are modeled in DESIRE (Brazier-Treur et al. 1997) in a meta-language, which is order sorted predicate logic. At BID the functional or logical relations between motivational attitudes and between motivational attitudes and informational attitudes are expressed as meta-knowledge, which may be used to perform meta-reasoning resulting in further conclusions about motivational attitudes. If we were to plan with BID with intelligent multimedia the logical relations might have to be amongst worlds forming the attitudes and event combinations. For example, in a simple instantiation of the BID model, beliefs can be inferred from meta-knowledge that any observed fact is a believed fact and that any fact communicated by a trustworthy agent is a believed fact. With IM_BID, the observed facts are believed facts only when a conjunction of certain worlds views and evens are in effect and physically logically visible to the windows in effect. Since planning with IM_BID is at times with the window visible agent groups, communicating, as two androids might, with facial gestures, for example Picard (1999a).

In virtual or the "real-world" AI epistemics, we have to note what the positivists had told us some years ago: the apparent necessary facts might be only tautologies and might not amount to anything to the point at the specifics. A BID assumption is that information is classified according to its source: internal information, observation, communication, deduction, assumption making. Information is explicitly labeled with these sources. Both informational attitudes (such as beliefs) and motivational attitudes (such as desires) depend on these sources of information. Explicit representations of the dependencies between attitudes and their sources are used when update or revision is required. A third assumption is that the dynamics of the processes involved are explicitly modeled. A fourth assumption is that the model presented below is generic, in the sense that the explicit meta-knowledge required to reason about motivational and informational attitudes has been left unspecified. To get specific models to a given application this knowledge has to be added. A fifth assumption is that intentions and commitments are defined with respect to both goals and plans. An agent accepts commitments towards himself as well as towards others (social commitments). For example, a model might be defined where an agent determines which goals it intends to fulfill, and commits to a selected subset of these goals. Similarly, an agent can determine which plans it intends to perform, and commits to a selected subset of these plans. Most reasoning about beliefs, desires, and intentions can be modeled as an essential part of the reasoning an agent needs to perform to control its own processes. The task of belief determination requires explicit meta-reasoning to generate beliefs. Desire determination: Desires can refer to a (desired) state of affairs in the world (and the other agents), but also to (desired) actions to be performed. Intention and commitment determination: Intended and committed goals and plans are determined by the component intention_and_commitment_determination. This component is decomposed into goal_determination and plan_determination. Each of these subcomponents first determines the intended goals and/or plans it wishes to pursue before committing to a specific goal and/or plan. Since the basic IM computing visual objects are hybrid pictures we define new planning techniques with hybrid pictures. The IM planning does not only applies planning with agents, it applies Morph Gentzen rules to hybrid pictures to achieve plan goals. The hybrid pictures carryout responsive, proactive, and reactive planning, only initiated and directed by a planning system. An example IM planning mission is as follows.

Hybrid picture 1—Spacecraft A Navigation Window

Agents: A1 Computes available docking times based on the visual field on the window.

A2 Carrys out docking sequence based on messages to Spacecraft B

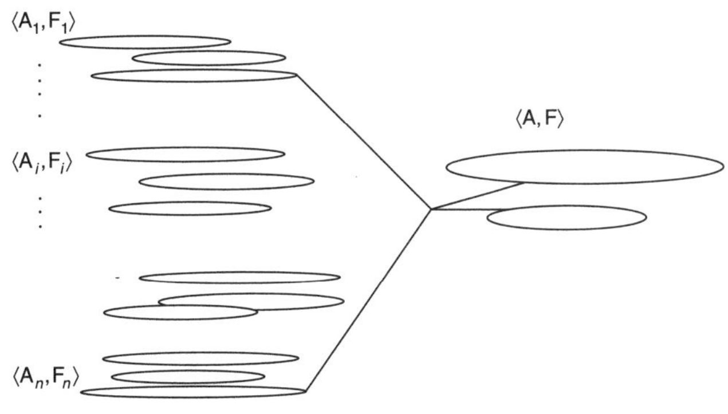

Figure 1. Multiagent object enterprise model.

Hybrid picture 2—Spacecraft B Navigation Window

Agents: B1 carries on course based on its visual field window

B2 Accepts and carries out docking maneuvers from external hovering craft agents.

Plan Goal Engage docking between A and B at appropriate A and B field windows. Morph Gentzen computing can be applied to the hybrid pictures to satisfy a plan goal. Thus morphing is applied with precise fluidity to plan computation. Proposition Morph Gentzen and Intelligent languages provide a sound and complete logical basis to VR.

3. THE SPATIAL MULTIAGENT NAVIGATOR

The autonomous space vehicles, e.g., Mars Rovers, are example areas where we have provided applications for spatial agent computing. Space examples are areas where there are specific terrains precomputed for missions. For such environments Morph Getnzen Spatial Logic can be designed to carry out autonomous intelligent multimedia activities. Morph Gentzen terrain logic is designed where a combinations known terrain events vision sensed trigger a specific autonomous activity by a Mars Rover in real-time. The microrover technology has several limitations precluding more ambitious science-rich missions. Onboard machine intelligence provides capabilities for autonomous search and recognition of potentially interesting targets, as well as capabilities for sensor platform planning and utilization. Morph Getnzen terrain logic can be applied to enhance autonomous traversal and autonomous multimedia search.

4. THE AUTONOMOUS ENTERPRISE

The designs state all exceptions to actions and what recovery and corrective actions are to be carried out. For each action on an object a dual action is to be supplied such that a specifier can fully define the effect of the dual actions. As an illustration the following trivial example defines part of a flight system's operations. Object := Enterprise_Flight_Computer; Set module at Locations Scanned by the Virtual_Moveer scanning device; Energise; Exp := OPS := Virtual_Move (Source, Destination, Object)|...; EVirtual_Move (Source, Destination, Object) |...; Virtual_Move (Mars, Enterprise, Module) ⇒ Signal an available Mars_Rover robot to Fetch Module from ⟨stored_location, module number⟩; EVirtual_Move(Mars, Enterprise, Module) ⇒ If Mars Rover robot signals that Module is locked or too heavy for robot to move to Virtual_Moveable points, Activate_Multiple_Robots to cooperate, unlock and move module to Virtual_Move point. An example model diagram function is Virtual_Move. A Morph Gentzen design might be applied to precomputed terrain multimedia chips. The pi's are processes. Semicolon represents processes sequencing. For example, $p_1; p_2$ represents processes p_1 and p_2 running sequentially, p_1 followed by p_2. Each of the message syntactic categories can be defined in very simple terms to suit the particular intended application areas. In addition, the specification language is extendible as far as the type of communication it allows the processes and the nature of parallelism. The process bodies can be specified in terms of the actions they are intended to perform and the messages they send.

$$\langle \text{module} \rangle := \{p_1; p_2; \ldots; p_n\} || \{q_1; q_2; \ldots; q_n\} || \langle \text{module} \rangle || \langle \{p_i\} \rangle$$

Developing robotic rovers for planetary surface exploration are areas where multiagent Morph Gentzen multimedia planning might be applied.

4.1. Plans and Tasks

The Planning Vocabulary we invoke is as follows.

Basic planning as in the introduction.

Domain Dependent—domain specific heuristics are applied to control planner operations.

Domain Independent—planners in which planning KR and algorithms are expected to work for a reasonably large variety of applications domains.
Task Level Planning—Redefined Tasks, Predefined Programs, Fixed Robot World View
Mission Planning—Generating activity plans from specific domain parameters for automating activities relevant to the mission.

4.2. The Planning Picture

Planning as Search: A potentially large search space for selecting a plan. Model views and heuristics can help. Conjunctive Goal Planning: Several simultaneous goals are tackled to have a marked effect on the search space. Plan Levels at Abstrips separated the goals by prioritizing importance. Hence abstract and general goals are satisfied before. The concrete and detail levels are treated next. Distributed and Agent Planning—Agents assigned to plan and applied distributed resources of problem-solving expertise. Metaplanning—planner reasons about goals and the techniques for generating plans. Model-Theoretic Planning: The World Model is defined form diagrams for models. Plans are operations resulting to terms at which goals are realized at the worlds. Domain Independent Planning—it can apply to many planning techniques, e.g., model-theoretic planning or distributed and multiagent planning. Action order plans—Represent plan as ordered actions; define points in the search space as partially elaborated plans; start form an initial state in the partial plan, apply a sequence of plan transformations to a fully detailed plan; Robot planning where robot's are situated in a world and acts based on an internal model for the external reality.

4.3. Practical Standard Planning Problems

Determining the truth of a formula at a particular point in a plan.
Deduction of context dependent effects.
Unification of two variables once they have constraints on them Handling parallel interactions Detecting and resolving Resource Conflicts.
Efficient search through the space of possible plans. The JPL Millennium project.
Planning with models—Nourani 1986–1997.
Intelligent Multimedia Spatial Planners—Nourani 1998–1999.

4.4. The Newer Planning Projects

Since the are further advance planning projects as indicated. Here. Plan Abstraction & Simplification-ala Fikes-Nilsson and Sacerdoti; Planning With Agents; Reactive Planning For robots; Hierarchical Planning; Practical Standard Planning Problems. How the standard planner is to interact with the preceding planner components is a very complex design process. The IM_BID model is applied with agents before presenting individual agent goals to the standard planner via the Agent Loop manager.[1]

4.5. Plans and Schedules for Spacecrafts

There are two types of schedules applied to planning. For spacecraft function plans there are events scheduled during the plan period. Scheduled evens are actions over which the planner has no control, in the sense of initiating them, but are known to occur at defined times. Plan activity schedules for plan process synchronization. Example new projects are planning and scheduling systems to generate spacecraft commands from high level goals in the new Millennium goal-driven command software projects at JPL (Chien et al. 1997). Heuristics are important in producing effective schedules. Chronologies are a way to get at the current state of volatile schedules. Chronologies are limited histories of the scheduling process. The planning-scheduling requirements for NASA deep space missions are extremely time and labor intensive. For the Voyager II encounter with Uranus the generation of acceptable schedules took work-decades of effort. The number of requested tasks far exceeds the

[1] The agent planner design and the multiboard platform are the author's venture properties. Applications must be with venture partnerships and proper credit.

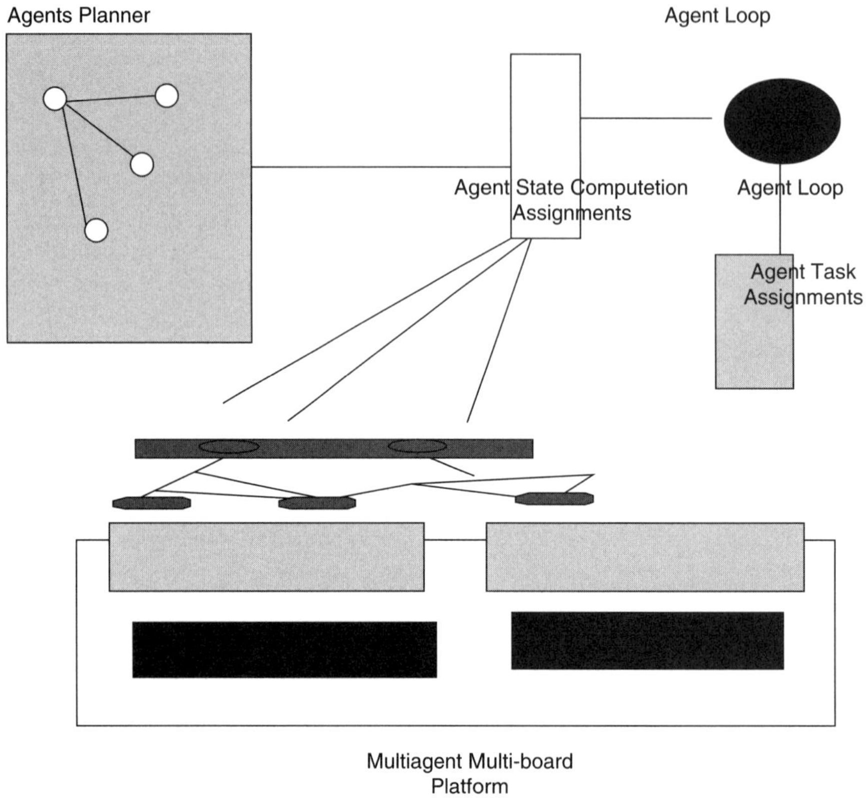

Figure 2. Agent planner with multiagent multi-board platform.

capabilities of the spacecraft leading to highly over-subscribed spacecraft resources. Further more, the domain constraints are insufficient to narrow down the scheduling choices. It is because the problem space is under-constrained. Chronology Directed Search Scheduling is applied to the Operations Mission Planner OMP at JPL in 1989.

4.6. Planning and Control Abstraction

Difficult scheduling problems were presented by the Space Station Freedom Project. Changing complements of payloads compete with each other and with the core system for resources. They potentially interfere with each other rights to an operating environment—a form of disturbing influences. Control Abstraction for planning scheduling might be defined to support the operators in establishing their schedule, and in responding to schedule changes in targets of opportunity. Provide an operating envelope schedule to allow users to develop detailed schedules. Limit requirements for the knowledge of individual payloads to the external effects permitted during or necessary for a control abstraction block. Control abstraction to for planning scheduling with constraints had been applied by Nourani since 1986. The Space Station Freedom Project reported the following techniques in 1991 for scheduling control abstraction: A planning scheduling technique is defined by separating the schedule integration from the detailed scheduling of individual operations. Many scheduling problems require constraints to be satisfied so that the schedules can be met, but do not require that the system reason about how the world changes as scheduled events occur. On the other hand, if the set of actions must be generated based on current situation and the current goals, then the problem solution itself also involves planning. For fixed location fixed view world robotics, e.g., a preprogrammed task specific robot often times task level programming and intelligent scheduling does most of the job. Since the problem is for the most part what ISIS sorts of systems are designed for. When it comes to autonomous spacecraft planning the latter applies. Furthermore, the problems are compounded with altering world views and schedules. For most AI problems the restrictions imposed by the problem constrain the admissible solutions to the extent that least commitment and constrain propagation techniques may converge to a solution schedule.

5. PLANNING AND SPATIAL NAVIGATION

Multiple AI worlds populated by agents having many concurrent goals present new planning problems. Neither a purely reactive nor standard planning might apply. The former does not apply because the agents have to know what they will be doing in the future to make decisions what to do in the present. The latter does not fully apply because of the size and the scope of the problems presented. Multilevel multiagent hybrid planning has to be defined. The project applies knowledge representation on generalized diagrams. The method of generalized diagrams, abbreviated by G-diagram, from Nourani (1983, 1995d) were invented for AI planning and reasoning, formulating various notions of generalized and free diagrams. It shows that G-diagrams from the basis for minimal efficient knowledge representation, henceforth abbreviated by KR, paradigms. Areas touched upon are foundational and formal. Essential characteristics of visual representations, diagram understanding and interpretation, properties of animated and changing diagrams, knowledge representation and inference, visual reasoning with diagrammatic programming languages, modelling interaction with diagrams, sound logical reasoning with diagrams, combination of diagrammatic knowledge and domain knowledge. Many examples are drawn from theories of AI to planning for robots to show the applicability of the techniques and theories proposed. World model revision is one of the difficult aspects of the nonmonotonic reasoning systems. Nourani (1983, 1991) present ways to get around it with generalized diagrams, where models are implicitly revised through dynamic changes to the diagrams from which the models are built. Hence we show that KR by G-diagrams simplifies the world revision problem. Generalized diagrams are used to build models with a minimal family of generalized Skolem functions. The minimal set of function symbols is functions with which a model can be built inductively. The functions can correspond to objects defining shapes and depicting pictures. We cannot formalize the real world. AI Worlds are relevant descriptions for problem solving in the real world parts we want models for G-diagrams can analogically model the semantics of a problem domain. The specific spatial areas addressed are cognitive structure of spatial knowledge, languages of spatial relations spatial and temporal reasoning, navigation in spatial environments, incomplete or imprecise spatial knowledge, presentation of spatial information Virtual reality and robot navigation.

6. VIRTUAL WORLDS, MODELS, AND PLANS

6.1. Relevant Worlds

Knowledge representation has two significant roles: to define a model for the AI world, and to provide a basis for reasoning techniques to get at implicit knowledge. An ordinary diagram is the set of atomic and negated atomic sentences that are true in a model. Generalized diagrams are diagrams definable by a minimal set of functions such that everything else in the model closure can be inferred, by a minimal set of terms defining the model. Thus providing a minimal characterization of models, and a minimal set of atomic sentences on which all other atomic sentences depend. Let us define a simple language $\mathscr{L} = \langle\{\text{tweety}\}, \{a\}, \{\text{bird}\}$, predicate letters, and FOL$\rangle$. A model may consist of {bird(tweety), ¬penguin(tweety) → bird(tweety), bird(tweety) v - bird(tweety), ...}. others may consist of $\{p(a), \neg p(a) \rightarrow p(a), p(a)\, v\, p(x), p(a)\, v\, p(x)\, v\, p(y), \ldots\}$. For AI we want models that could be computed effectively and efficiently. Thus, it is useful to restrict the types of models that we define for real world applications. Primarily, we are interested in models with computable properties definable from the theory. In order to point out the use of the generalized method of diagrams we present a brief view of the problem of planning from Nourani (1995d) within the present formulation. The diagram of a structure in the standard model-theoretic sense is the set of atomic and negated atomic sentences that are true in the structure. The generalized diagram (G-diagram) Nourani (1991, 1995d) is a diagram in which the elements of the structure are all represented by a minimal specified set of function symbols and constants. Thus it is sufficient to define the truth of formulas only for the terms generated by the minimal family of functions and constant symbols. Such assignment implicitly defines the diagram. This allows us to define a canonical model of a theory in terms of a minimal family function symbols. By definition a diagram is a set of atomic and negated atomic sentences, and can therefore be considered as a basis for defining a model, provided we could by algebraic extension, define the truth value of arbitrary formulas instantiated with arbitrary terms. Thus all compound sentences build out of atomic sentences then could be assigned a truth value, handing over a model. The following examples would run throughout the paper Consider the primitive first order language (FOL) $\mathscr{L} = \{c\}, \{f(X)\}, \{p(X), q(X)\}$. Let us apply Prolog notation convention for constants and variables and the simple theory {for all X: $p(X) \rightarrow q(X), p(c)$}, and indicate what is meant by the various notions.

$$\begin{aligned}(\text{model}) = \{&p(c), q(c), q(f(c)), q(f(f(c))), \ldots\},\\ \{&p(c)\&q(c), p(c)\&p(X), p(c)\&p(f(X)), \ldots\},\\ \{&p(c)\, v\, p(X), p(c)\, v\, p(f(X)), p(c) \rightarrow \neg p(c) \ldots\}\end{aligned}$$

(diagram) $p(c), q(c), p(c), q(f(c)), q(f(f(c))), \ldots\}, \ldots, q(X)\}$, i.e., diagram = the set of atomic formulas of a model. The diagram is (diagram) = $\{p(c), q(c), q(f(c)), q(f(f(c))), \ldots, q(X)\}$. There are various notions of diagram from Nourani's earlier papers (see references) that are applied in the present paper are presented here such that we can apply them. The term generalized is applied to indicate that such diagrams are defined by from basic terms and constants of a language to fully define diagrams making use of only a minimal set of functions. The G-diagrams are applicable to KR for planning with incomplete knowledge (Nourani 1995d) and free proof trees (Nourani 1995d, Nourani and Hoppe 1993). Plausible diagrams are diagrams with formulas that get their truth assignment based on a belief function that is defined from a degree of plausibility assigned to each formula on the diagrams. There were applications by the author (Nourani 1991) to represent plausible knowledge for planning applications. Predictive diagrams were presented since Nourani (1995d), Nourani and Hoppe (1993) and apply it for KR to provide a model-theoretic characterization for planning where there is a predictive function on incomplete knowledge.

6.2. Virtual Trees

In our projects present paper we also instantiate proof tree leaves with free Skolemized trees. Thus virtual trees are substituted for the leaves. In the present approach, as we shall further define, leaves could be virtual trees. By a virtual tree we intend a term made of constant symbols and Skolem functions terms A plan is a sequence of operations in the universe that could result in terms that instantiate the truth of the goal formulas in the universe. That's what goes on as far as the algebra of the model is concerned. It is a new view of planning prompted by our method of planning with GF-diagrams and free Skolemized trees. It is a model-theoretic view. Proof-theoretically a plan is the sequence of proof steps (Fikes and Nilsson 1971) that yields the proof for the goal formula. The proof theoretic view is what the usual AI literature presents. The planning process at each stage can make use of GF-diagrams by taking the free interpretation, as tree-rewrite computations (for example, Nourani 1996a) of the possible proof trees that correspond to each goal satisfiability. The techniques we have applied are to make use of the free Skolemized proof trees in representing plans in terms of generalized Skolem functions. In planning with GF-diagrams that part of the plan that involves free Skolemized trees is carried along with the proof tree for a plan goal. We can apply predictive diagram KR to compute queries and discover data knowledge from observed data and visual object images keyed with diagram functions. Model-based computing (Nourani 1995d) which can be applied to automated data and knowledge engineering with keyed diagrams. Specific computations can be carried out with predictive diagrams (Nourani 1995d). For cognition, planning, and learning the robot's mind a diagram grid can define state. The starting space applicable (Nourani 1991) project was meant for autonomous robots wandering at outer space. The designs in Nourani (1991, 1994) are ways for a robot to update its mind state based on what it encountered on its path. What the robot believes can be defined on a diagram grid. The degree to which a robot believes something is on the grid. It can get strengthened or weakened as a function of what the robot learns as progress is brought on. Robot's Mind State—The array grid entries are pointing to things to remember and the degree the robot believes them. The entry 15 is an item robot believes the most. Logically and theoretically the grid is minimally defined by the G-diagram functions. The gird model is a way to encode diagrammatic reasoning.

6.3. Morph Gentzen Virtual Reality

The IM Morphed Computing Logic for computing for multimedia is new projects with important computing applications since Nourani (1998b, d). The basic principles are a mathematical logic where a Gentzen (1943) or natural deduction (Prawitz 1973) systems is defined by taking arbitrary structures and multimedia objects coded by diagram functions. Multimedia objects are viewed as syntactic objects defined by functions, to which the deductive system is applied. Thus we define syntactic morphings (Nourani 1998b) to be a technique by which multimedia objects and hybrid pictures are morphed and, perhaps holographically, mapped via their defining functions to a new hybrid picture. Functorial topological structures can be defined without difficulty. The deduction rules are a Gentzen system augmented by Morphing, and Transmorphing. The logical language has function names for hybrid pictures. The MIM Morph Rule—An object defined by the functional n-tuple $\langle f_1, \ldots, f_n \rangle$ can be Morphed to an object defined by the functional n-tuple $\langle h(f_1), \ldots, h(f_n) \rangle$, provided h is a homomorphism of abstract signature structures Nourani (1993c). The MIM Trans-Morph Rules—A set of rules whereby combining hybrid pictures $p_1, \ldots, p_n$ defines an Event $\{p_1, p_2, \ldots, p_n\}$ with a consequent hybrid picture p. Thus the combination is an impetus event. By transmorphing hybrid picture's corresponding functions a new hybrid picture is deduced. The techniques can be applied to arbitrary topological structures. The languages and MIM rules are applied to algebraic structures. The deductive theory is a Gentzen system in which hybrid pictures are named by parameterized functions; augmented by the MIM Morph and transform rules. The Model theory is defined from Intelligent syntax languages (Nourani 1995f, 1996a). A computational logic for intelligent languages is presented in brief with a soundness and completeness theorem in Nourani (1996a). The idea is to do it at abstract models syntax trees without specifics for the shapes and topologies applied. We can start with $\mathscr{L}_{\omega_1, \omega}$ and further on apply well-behaved infinitary languages. A soundness and completeness theorem has been put forth Nourani (1998b) stating the MIM Soundness and Completeness: Theorem Nourani (1997) Morph Gentzen Logic is sound and complete (see Chapter 6).

6.4. Intelligent Trees and Spatial KR

Visual objects connected by agents carrying information represent the visual field amongst objects about the field, and carried onto intelligent trees for computation. Intelligent trees compute the spatial field information with the diagram functions. The trees defined have function names corresponding to computing agents. The computing agent functions have a specified module defining their functionality. Intelligent trees are applied to compute field information with agents where all computation is expressed and carried on intelligent languages (Nourani 1996a). By an intelligent language we intend a language with syntactic constructs that allow function symbols and corresponding objects, such that the function symbols are implemented by computing agents. A set of function symbols in the language, referred to by AF, is the set modeled in the computing world by AI Agents with across and/or over board capability. The boards, message passing actions, and implementing agents are defined by syntactic constructs, with agents appearing as functions. The computation is expressed by an abstract language that is capable of specifying modules, agents, and their communications.

7. SPATIAL MORPH GENTZEN

Multiagent spatial vision techniques are introduced in the paper. The duality for our problem solving paradigm (Nourani 1991, 1993c) is generalized to be symmetric by the present paper to formulate Double Vision Computing. The basic technique is that of viewing the world as many possible worlds with agents at each world that compliment one another in problem solving by cooperating. The author presented an asymmetric view of the application of this computing paradigm and the basic techniques were proposed for various AI systems (Nourani 1991). The double vision computing paradigm with objects and agents might be depicted by the following figure. For computer vision (Winston 1975), the duality has obvious anthropomorphic parallels. The object coobject pairs and agents solve problems on boards by cooperating agents. The cooperative problem solving paradigms have been applied ever since the AI methods put forth (Hays-Roth 1985). However, the multiagent multiboard techniques due to the author (1995a) Figure1-lower boards depict the Multiagent Multiboard Spatial Vision design. The techniques to be presented are to be applied to Mobile Multimedia. Communication and computation by multimedia visual-object languages can be programmed with IM with a simple syntax. The techniques to be presented are to be applied for (a) Precomputed video-object composition and combination for spatial Morph Gentzen computing with visual objects (b) High speed visual spacecraft navigation by multiagent multimedia.

7.1. The Spatial Multiagent Navigator

The autonomous space vehicles, e.g., Mars Rovers, are example areas where we have provided applications for spatial agent computing. Space examples are areas where there are specific terrains precomputed for missions. For such environments Morph Getnzen Spatial Logic can be designed to carry out autonomous intelligent multimedia (Nourani 1998d) activities. MIM terrain logic is designed where a combinations known terrain events vision sensed engage a specific autonomous activity by a Mars Rover in real-time. Chipcraft is a science and engineering technique to design general purpose or application specific multimedia chips at the IV-98, Stuttgart. NASA-JPL is developing technologies that enable 20-kilogram-class microrovers to autonomously traverse many kilometers on the surface of Mars, perform scientist-directed experiments, and return relevant data back to Earth. Present microrover technology has several limitations precluding more ambitious science-rich missions. Onboard machine intelligence provides capabilities for autonomous search and recognition of potentially interesting targets, as well as capabilities for sensor platform planning and utilization. Morph Getnzen terrain logic can be applied to enhance autonomous traversal and autonomous multimedia search.

7.2. Multiagent Visual Planning

The balloons are visual objects, the squares agents, the dotted lines the message paths. Multiagent spatial vision techniques are introduced in Nourani (1998a, b). The duality for our problem solving paradigm Nourani (1995a, b) is generalized to be symmetric by the present paper to formulate Double Vision Computing. The basic technique is that of viewing the world as many possible worlds with agents at each world that compliment one another in problem solving by cooperating. The author presented an asymmetric view of the application of this computing paradigm and the basic techniques were proposed for various AI systems (Nourani 1991). The double vision computing paradigm with objects and agents might be depicted by the following figure. For computer vision (Winston 1975), the duality has obvious anthropomorphic parallels. The object co-object pairs and agents solve problems on boards by co-operating agents.

7.3. The Intelligent Multimedia Beliefs, Intentions, Desire Model

The co-operative problem solving paradigms have been applied ever since the AI methods put forth by Hays-Roth (1985). The multiagent multi-board techniques are due to the author (1995a). The BID [10] model has to be enhanced to be applicable to intelligent multimedia.

Let us start with an example multi-board model where there multiagent computations based on many boards, where the boards corresponds to either virtual possible worlds or to alternate visual views to the world, or to the knowledge and active databases.

EXERCISES

1. Specify a multiagent autonomous flight control system for a known spacecraft.
2. Design visual navigation with agents and visual field boards.
3. Design some specific Morph Gentzen based window field visual navigation based on known spatial KR.

REFERENCES

1. M. Bienkowski, *IEEE Expert* 10, 27 (1995).
2. F. M. T. Brazier, B. Dunin-Keplicz, N. R. Jennings, and J. Treur, "DESIRE: modelling multi-agent systems in a compositional formal framework," (M. Huhns and M. Singh, Eds.), Vol. 1. International Journal of Cooperative Information Systems, 1997.
3. B. Chandrasekaran, Janice Galsgow, and N. H. Narayanan, Eds., "Diagrammatic Reasoning." 1996.
4. S. Chien, D. DeCosta, R. J. Doyle, and P. Stolorz, *AI Magazine* 103 (1997).
5. R. E. Fikes and N. J. Nilsson, *Artificial Intelligence* 2, 189 (1971).
6. M. Genesereth and N. J. Nilsson, "Logical Foundations of Artificial Intelligence," Morgan-Kaufmann, Los Altos, 1987.
7. G. Gentzen, *Math. Ann.* 119, 140 (1943).
8. M. Gini and G. Gini, "Towards Automatic Recovery in Robot Programs," Vol. 2, pp. 821–823. Proc. 8th IJCAI, Karlsruhe, Germany, 1983.
9. B. Hays-Roth, *Artificial Intelligence* 26, 251 (1985).
10. J. P. Muller and M. Pischel, "Modeling Interactive Agents in Dynamic Environments," Proceedings 11th European Conference on AI, John Wiley and Sons, Amsterdam, The Netherlands, 1994.
11. F. Northrup III, S. E. Cross, and C. Owens, *IEEE Expert* 10, 4 (1995).
12. C. F. Nourani, "Equational Intensity, Initial Models, and AI Reasoning," Technical Report: A Conceptual Overview, in Proc. Sixth European Conference in Artificial Intelligence, Pisa, North-Holland, Italy, 1983.
13. C. F. Nourani, "Planning and Plausible Reasoning in AI," pp. 150–157. Proc. Scandinavian Conference in AI, Roskilde, Denmark, IOS Press, 1991.
14. C. F. Nourani, "Abstract Implementations By Computing Agents: A Conceptual Overview," Proc. SERF-93, Orlando, FL, 1993c.
15. C. F. Nourani, "Towards Computational Epistemology—A Forward, Summer Logic Colloquium," Clairemont-Ferrand, France, 1994.
16. C. F. Nourani, "Double Vision Computing." IAS-4, Intelligent Autonomous Systems, Karlsruhe, Germany, 1995a.
17. C. F. Nourani, "AII and Heterogenous Software Design," MAAMAW'97, Eighth European Workshop on Modelling Autonomous Agents in A Multi-Agent World, University of Karlskrona/Ronneby, Ronneby, Sweden, 1995b.
18. C. F. Nourani, "Multiagent Robot Supervision," ECML Learning Robots, Heraklion, 1995c.
19. C. F. Nourani, "Free Proof Trees and Model-Theoretic Planning." AISB, Sheffiled, 1995d.
20. C. F. Nourani, "Visual Computational Linguistics—The Preliminary Definition." 34th International Linguistics Colloquium, Mainz, Germany, 1995e.
21. C. F. Nourani, "Automatic Models From Syntax," Scandinavian Linguistics, Oslo, Norway, 1995f.
22. C. F. Nourani, "Slalom Tree Computing," AI Communications, IOS Press, Amsterdam, 1996a.
23. C. F. Nourani, "A Parallel Functional Modular Language With Agents and Distributed Objects," Brief, IFL Wkshp Parallel Functional Languages, London, 1998a.
24. C. F. Nourani, "MIM Logik," Summer Logic Colloquium, BSL, Symbolic Logic Publications, Prague, 1998b.
25. C. F. Nourani, "Intelligent Trees, Thought Models, And Intelligent Discovery Model-Based Reasoning in Scientific Discovery (MBR'98)," pp. 17–19. Pavia, Italy, 1998c.
26. C. F. Nourani, "Intelligent Multimedia—New Computing Techniques and Its Applications." CSIT, Moscow, 1998d.
27. C. F. Nourani, "Double Vision Spacecraft and Autonomous Flight Control." AA99 Autonomy Control Track, Seatlle, WA, 1999a.
28. C. F. Nourani and T. Hoppe, "GF Diagrams, and Free Proof Trees." Technical Report, Berlin Logic Colloquium, Potsdam, 1993.
29. R. W. Picard, "Affective Computing for HCI, Proceedings of HCI," Munich, Germany, 1999a.
30. D. Prawitz, "Natural Deduction: A Proof Theoretic Study." Stokholm, Almqvist and Wiksell, 1973.
31. E. Sacerdti, *Artificial Intelligence* 5, 115 (1974).
32. S. A. Vere, *Aerspace America* 50, 1983.
33. D. Wilkins, *Artificial Intelligence* 22, 269 (1984).
34. P. H. Winston, "The Psychology of Computer Vision" (McGraw Hill, Ed.). New York, 1975.

CHAPTER 10

Knowledge Representation and Predictive Intelligent Multimedia Diagram Models

CONTENTS

1. INTRODUCTION

The computational models for what we have seen in the preceding chapters have been abstract at times. For some application areas, the specific logic computations are necessary. This chapter is a basis for model discovery, computation, KR, and model checking on

ISBN: 1-58883-037-3/$35.00

Intelligent Multimedia Computing Science
By Cyrus F. Nourani
Pages: 123–136

generalized diagrams with automated deduction. The applicability of G-diagram method for KR to partial deduction and abduction is presented in brief, where we have defined abductive diagrams. Furthermore, we make a brief connection to KR for proof abstraction methods in AI, to show the applicability of the methods we are presenting with free proof trees.

Many examples are drawn from theories of AI to planning for robots to show the applicability of the techniques and theories proposed. World model revision is one of the difficult aspects of the nonmonotonic reasoning systems. Our papers (Nourani 1984, 1991) presented solution to it by the generalized diagram formulation, where models are implicitly revised through dynamic changes to the diagrams from which the models are built. Thus, we show that KR by G-diagrams simplifies the world revision problem. The formulation of default in the author's papers, as further applied to abductive reasoning in this paper is referring to model-theoretic consistency. The practical problem of incomplete information in AI reasoning systems requires model revision when new information is given about the system. It is precisely for this reason that we argue for methods, which build models directly from the theory—incrementally and perhaps automatically, such that theory revision can be accommodated automatically. Truth maintenance and model revision can all be done by a simple reassignment to the G-diagram. The canonical model of the world is defined directly from the diagram.

Generalized diagrams are used to build models with a minimal family of generalized Skolem functions. The minimal sets of function symbols are functions with which a model can be built inductively. The functions can correspond to objects defining shapes and depicting pictures. We cannot formalize the real world. AI *Worlds* are relevant descriptions for problem solving in the real world parts. The areas where we have applied G-diagrams include formal theories of Diagrammatic Reasoning. Descriptive computing with generic diagrams is presented with applications to model computing and discovery. The representations for reaching an object might be infinitary only. In Chapter 5, we saw instances that the artificial intelligence problem from the robot's standpoint is to acquire a decidable descriptive computation for the problem domain. The mathematical specifics are stated here. Further, Boolean computing bases to models are stated with specific mathematical foundations.

2. KNOWLEDGE REPRESENTATION AND INTELLIGENT MULTIMEDIA SPECIFICS

Knowledge representation has two significant roles: to define a model for the AI world, and to provide a basis for reasoning techniques to get at implicit knowledge. An ordinary diagram is the set of atomic and negated atomic sentences that are true in a model. Generalized diagrams are diagrams definable by a minimal set of functions such that everything else in the models closure can be inferred, by a minimal set of terms defining the model. Thus providing a minimal characterization of models, and a minimal set of atomic sentences on which all other atomic sentences depend.

We want to solve real world problems in AI. Obviously, for automating problem solving, we need to represent the real world. Since we cannot represent all aspects of a real world problem, we need to restrict the representation to only the relevant aspects of the real world we are interested in. Let us call this subset of relevant real world aspects the *Relevant World* for a problem. AI approaches to problem solving represent the knowledge usually in some kind of first-order language, consisting of at least constants, functions and predicate symbols.

Our primary focus will be the relations amongst KR, AI worlds, and the computability of models. Truth is a notion that can have dynamic properties. A formalization of problem solving knowledge (we refer to it from now on with the term: theory) is not independently true. Neither is a consequence inferred from a theory true independently. The sets of sentences defining a theory are only true with respect to a certain world. This insight has found its theoretical counter-part in model theory, which fixes the semantics of a language (most often First Order Logic) with respect to a certain world by means of an interpretation function.

Interpretation functions map language constructs (constants, function and predicate symbols) onto entities of the world, and determines the notion of truth for individuals, functions and relations in the domain.

The real world is infinite as are AI worlds at times. We might be interested to figure out in which AI worlds a theory or a sentence will be valid. Furthermore, we might like to perform abstract inferences over equivalence classes of models (an issue of precomputation and reasoning). We have to be able to represent these ideas with computable formulations. Another instance is if we need to revise a robot's beliefs about an AI world. There is also the instance where we could be interested in the different ways formalization can be interpreted with respect to a certain AI world (an issue of user modeling). Thus, we are confronted with a huge number of possible truth assignments. We usually have to cope in even finite AI worlds with an exponential number of possible truth assignments. Keeping the models, which need to be considered small and to keep a problem tractable, such that the models could be computable and within our reach, are important aspects. Related questions are also taken up in another areas of research by the first author in Nourani (1991, 1993). We show that we can apply G-diagram functions to localize reasoning to the worlds affected by some relevant functions to a specific reasoning aspect.

3. COMPUTABLE AI WORLD MODELS

To prove Godel's completeness theorem, Henkin defined a model directly from the syntax of the given theory. This structure is obtained by putting terms that are provably equal into equivalence classes, then defining a free structure on the equivalence classes. The reasoning enterprise requires more general techniques of model construction and extension, since it has to accommodate dynamically changing world descriptions and theories. The techniques from Nourani (1983) for model building as applied to the problem of AI reasoning allows us to build and extend models through diagrams. This requires us to define the notion of generalized diagram. The diagrams are used to build models with a minimal family of generalized Skolem functions. The minimal sets of function symbols are those with which a model can be built inductively. We focus our attention on such models, since they are Initial (Goguen 1973) and computable (Nourani 1980, 1983). The G-diagram methods applied and further developed here allows us to formulate AI world descriptions, theories, and models in a minimal computable manner. It further allows us to view the world from only the relevant functions. Thus, models and proofs for AI problems can be characterized by models computable by a set of functions.

3.1. Relevant Worlds and KR

Instead of using as a semantic basis for a theory, some pure mathematical structure like the Herbrand universe, Kripke structures, etc. we apply the method of generalized diagrams. We are interested in a semantic foundation of our methods and approaches which is based on the real world consisting of people, robots, languages, and, words etc. The usual Tarskian semantics allows us to use some symbols for denoting objects of the real world, to draw some inferences on those symbols and to obtain some statements that will then be true in the real world. Clearly, this is the basis for the symbolic computation paradigm of AI. It is possible, as we show by this paper and Nourani (1991, 1993), to define new symbolic computation aradigms for KR and AI reasoning based on G-diagrams, that have appealing computing properties. Hence, usually during modeling we focus only on some subparts of the real world, using only problem-relevant statements. What we do not know on a generalized diagram is defined in terms of a generalized Skolem function. We like to call such a restriction of the real world the Relevant World. Clearly, even such a restricted AI world may in some cases be still complex and infinite. However by such a restriction, we have already made the number of possible interpretations and thus the semantics of formalization considerably smaller.

3.2. Representing World Models by G-diagrams

An AI world consists of individuals, functions on them, and relations between them. These entities allow us to fix the semantics of a language for representing theories about AI worlds. We take the usual model-theoretical route, and assign via an interpretation function individuals to constants, functions to function symbols and relations to predicate symbols. Let us define a simple language $\mathscr{L} = \langle\{\text{tweety}\},\{a\},\{\text{bird}\}$, predicate letters, and FOL$\rangle$.

A model may consist of {bird(tweety), penguin(tweety) $\rightarrow$ bird(tweety), bird(tweety) $v\,\neg$ bird(tweety), . . . }, others may consist of

$$\{p(a), \neg p(a) \rightarrow p(a), p(a)vp(x), p(a)vp(x)vp(y), \ldots\}.$$

Because we can apply arbitrary interpretation functions for mapping language constructs into AI worlds, the number of models for a language is infinite. Although this makes perfect sense from a theoretical and logical point of view, from a practical point of view, this notion of model is too general for AI applications. Since for AI we want models that could be computed effectively and efficiently. Thus, it is useful to restrict the types of models that we define for real world applications. Primarily, we are interested in models with computable properties definable from the a theory.

3.3. KR on Diagrams and Partial Deductions

Proofs can be abstracted by generalizing away from constants in the proof. Thus, such a generalized proof can be defined by a whole class of minimal diagrams. This process is usually realized via partial deduction, which can be regarded as the proof-theoretical way of abducting diagrams whose literals are necessary conditions for the proof.

Under certain restrictions partial deduction can be transformed into abduction Hoppe (1992). We want to present a formal relation between partial deduction and abduction from a model-theoretical point of view. However, it was not clear how PD can be given a model-theoretical semantics and how knowledge is to be represented to a proof system. This is one reason why the formulation of nonmonotonic reasoning with G-diagrams presented by Nourani (1983) could be applicable here.

Let us now view the deductive methods. The proof-theoretic example is SLDNF-resolution, which is a well-known deductive mechanism (Remark: This makes it easier for us to link diagrams to PD and Abductive Reasoning). A SLDNF-proof can be considered as the unfolding of an AND/OR-tree, which is rooted in the formula to be proven, whose branches are determined by formulas of the theory, and whose leaves are determined by atomic formulas which are true in an AI world. In the present approach, as we shall further define, leaves could be free Skolemized trees. By a free Skolemized tree we intend a term made of constant symbols and Skolem function terms. Like models and diagrams, which where generalized above in different ways, we can generalize the notion of a proof. By dropping the assumption that proof-tree leaves get instantiated with atomic formulas, we get a more general notion of a proof, which is usually called "partial deduction" (Hoppe 1992). The proof methods are driven form knowledge represented on G-diagrams.

Partial deduction usually computes from a formula and a theory an existential quantified diagram. In the present paper we also instantiate proof tree leaves with free Skolemized trees, where free trees are substituted for the leaves. By a free Skolemized tree we intend a term built with constant symbols and Skolem functions terms. Dropping the assumption that proof-tree leaves get instantiated with atomic formulas only yields an abstract and general notion of proof trees. The mathematical formalization that allows us to apply the method of free proof trees is based on the authors projects since 1993, further developed and applied to AI here.

4. GENERALIZED DIAGRAMS AND RELEVANT WORLDS

4.1. Generalized Diagrams

There are various notions of diagram from Nourani's earlier papers (see references) that are applied in the present paper are presented here such that we can apply them. The term generalized diagram in this and other papers of first author referenced refers to diagrams that are instantiated with generalized Skolem functions. The author had defined generalized Skolem functions as Skolem functions with which initial models are defined inductively. In order to point out the use of the generalized method of diagrams we present a brief view of the problem of planning from the author's earlier projects within the present formulation. The diagram of a structure in the standard model-theoretic sense is the set of atomic and negated atomic sentences that are true in that structure. The generalized diagram (G-diagram) is a diagram in which the elements of the structure are all represented by a minimal family of function symbols and constants. Hence, it is sufficient to define the truth of formulas only for the terms generated by the minimal family of functions and constant symbols. Such assignment implicitly defines the diagram. This allows us to define a canonical model of a theory in terms of a minimal family function symbols.

DEFINITION 4.1. Let M be a structure for a language $\mathscr{L}$, call a subset X of M a generating set for M if no proper substructure of M contains X, i.e. if M is the closure of $X \cup \{c(M): c$ is a constant symbol of $\mathscr{L}\}$. An assignment of and $G: A \rightarrow M$, such that $\{G(a): a \in A\}$ is a set of generators for M. Interpreting a by $g(a)$, every element of M is denoted by at least one closed term of $\mathscr{L}(A)$. For a fixed assignment $\langle A, G\rangle$ of constants to M, the *diagram of* M, $D\langle A, G\rangle(M)$ is the set of basic (atomic and negated atomic) sentences of $\mathscr{L}(A)$ true in M. (Note that $\mathscr{L}(A)$ is $\mathscr{L}$ enriched with set A of constant symbols.)

DEFINITION 4.2. A G-diagram for a structure M is a diagram $D\langle A, G\rangle$, such that the G in Definition 4.1 has a proper definition by a specific function set.

The term *generic diagram* is at times applied in the authors publications to refer to the above as an alternate terminology to generalized diagram. Models uphold to a deductive closure of the axioms and some rules of inference as specified. By the definition of a diagram they are a set of atomic and negated atomic sentences. Thus the diagram might be considered as a basis for a model, provided we can by algebraic extension, define the truth value of arbitrary formulas instantiated with arbitrary terms. Thus, all compound sentences build out of atomic sentences then could be assigned a truth value, handing over a model. It will be made clearer in the following subsections. The following examples would run throughout the paper. Consider the primitive first order language (FOL)

$$\mathscr{L} = \{c\}, \{f(X)\}, \{p(X), q(X)\}$$

Let us apply Prolog notation convention for constants and variables) and the simple theory $\{$for all X: $p(X) \rightarrow q(X), p(c)\}$, and indicate what is meant by the various notions.

$$\begin{aligned}(\text{model}) = \{&p(c), q(c), q(f(c)), q(f(f(c))), \ldots\},\\ &\{p(c) \wedge q(c), p(c) \wedge p(X), p(c)\&p(f(X)), \ldots\},\\ &\{p(c)vp(X), p(c)vp(f(X)), p(c) \rightarrow \neg p(c) \ldots\}\end{aligned}$$

$$(\text{diagram}) = \{p(c), q(c), p(c), q(f(c)), q(f(f(c))), \ldots\}, \ldots, q(X)\}$$

i.e. diagram = the set of atomic formulas of a model. Thus the diagram is

$$(\text{diagram}) = \{p(c), q(c), q(f(c)), q(f(f(c))), \ldots, q(X)\}.$$

Based on the above, we can define generalized diagrams. The term generalized is applied to indicate that such diagrams are defined by algebraic extension from basic terms and constants of a language. The fully defined diagrams make use of only a minimal set of functions. Generalized diagram is (generalized diagram) $= \{p(c), q(c), p(f(t)), q(f(t))\}$ for t defined by induction, as $\{t_0 = c$, and $t_n = \{f(t_{n-1})\}$ for $n > 0$.

It is thus not necessary to redefine all $f(X)$'s since they are instantiated. Nondeterministic diagrams are those in which some formulas are assigned an indeterminate symbol, neither true nor false, that can be arbitrarily assigned in due time.

$$(\text{nondeterministic diagram}) = \{p(c), q(c), p(f(t)), q(f(c)), q(f(f(c))), I_q(f(s)))\},$$

t is as defined by induction before and $I_q(f(s)) = I_q$ for some indeterminate symbol I_q, for $\{s = t_n, n >= 2$. These G-diagrams are applicable for KR in planning with incomplete knowledge with backtracking. *Free Skolemized diagrams* are those in which instead of an indeterminate symbols there are free Skolem functions, that could be universally quantified.

$$\text{(free Skolemized diagram)} = \{p(c), q(c), p(f(t)), q(f(c)), q(f(f(c))), q_{\text{F}\cdot s}(s)\},$$

where t and s are as defined in the sections before. These G-diagrams are applicable to KR for planning with incomplete knowledge (Nourani 1994) and free proof trees. *Plausible diagrams* are diagrams with formulas that get their truth assignment based on a belief function that is defined from a degree of plausibility assigned to each formula on the diagrams. There were applied by the author (Nourani 1991) to represent plausible knowledge for planning applications.

A *generalized free diagram* (GF-diagram) is a diagram that is defined by algebraic extension form a minimal set of function symbols.

$$\text{(generalized free diagram)} = \{p(c), q(c), p(f(t)), q(f(t)), q_{\text{F}\cdot s}(s)\}$$

for t and s as before. A generalized free plausible diagram, is a GFD that has plausibility degrees assigned to it. These G-diagrams are applied for KR to planning with free proof trees (Nourani 1994).

4.2. Diagrams for World Computing

Existential quantified diagrams carry one main deficit. The Skolem objects are not characterized, Hilbert's epsilon symbol may resolve this. This not only allows us to make the connection between PD and abduction even closer, because we can abduce in a FOL (first order logic) framework the description of necessarily existing objects for making a proof succeed, it also allows us to characterize those objects. In the author's projects since 1994 these issues are dealt with for free proof trees. AI planning problems are further reviewed in Chapter 13 and the blocks world problems solver is defined there.

4.3. Diagrams and Incomplete KR

In this section, we extend the notion of generalized diagram (G-diagram) to include plausibility and nondeterminism for planning and for representation of possible worlds. An extended notion of G-diagram can encode possible worlds to capture the "maximally complete" idea and can be used for model revision and reconstruction. By assigning a plausibility ranking to formulas one can set a truth limit ordinal t as the truth threshold. These notions of diagram are applied, by way of example, to planning such that the notions of computations with diagrams and free proof trees can be illustrated. A nondeterministic diagram is a diagram with indeterminate symbols instead of truth values for certain formulas. For example,

$$\text{(nondeterministic diagram)} = \{p(c), q(c), p(f(t)), q(f(c)), q(f(f(c))), I_q(f(s)))\},$$

t is as defined by induction before and $I_q(f(s)) = I_q$ for some indeterminate symbol I_q, for $\{s = t \text{ sub } n, n \geq 2\}$. p and q are language predicate letters and l_q is a variable letter designated by the index letter q to indicate Boolean model assignments, e.g., $\{\text{T}, \text{F}\}$ only. Formulas with plausibility ranking less than t would be assigned 'T' and the other formulae would be assigned 'F.' Thus (10) defined the notion of a plausible diagram, which can be constructed to define plausible models for revised theories. In practice, one may envision planning with plausible diagrams such that certain propositions are deliberately left indeterminate to allow flexibility in planning. In (10) nondeterministic diagrams were defined by assigning an undefined "X" symbol to predicates in the diagram whose truth-values are not known at each stage of planning. Such extensions to the usual notion of diagram in model theory are put forth in the author's publications since 1995. That approach was one method of avoiding the computational complexity and computability problems of having complete diagrams. Truth maintenance and model revision can all be done by a simple reassignment to the diagram. The canonical model of the world is defined directly from the diagram.

4.4. Generalized Free Skolem Diagrams

Such diagrams are defined by assigning truth-values to predicates with known truth-values at each stage and generalized Skolem functions (Nourani-Hoppe 1995, Nourani 1991, 1995) to represent those predicates whose truth-values are not known. Thus, we define diagrams in which predicates are replace by their corresponding Skolem functions. For example in the block world example of AI, the entry in the diagram function for top (A, B) predicate is replaced by a Skolem function F_{top} (A, B). We refer to such diagrams as GF-Diagrams. The example from the model of sections above is as follows.

$$\text{(free Skolemized diagram)} = \{p(c), q(c), p(f(t)), q(f(c)), q(f(f(c))), q_F \cdot s(s)\},$$

where t and s are as defined in the sections before. In the present paper, we devise new methods of achieving nondeterminism by proposing what we shall refer to as Free Skolemized Diagrams. Note that the GF-diagrams, as in the approach by Nourani (1995), also include all the information required to define possible worlds. Absolute formulas can be assigned plausibility ranking which represent their truth-value through the belief function limit ordinal. Relativized truth is captured by altering the truth limit ordinal, and by leaving plausibility degrees free by assigning generalized Skolem functions for the unknown predicates.

4.5. Predictive Diagrams for KR

A *generalized predictive diagram* is a predictive diagram with D[M] defined from a minimal set of functions. The predictive diagram could be minimally represented by a set of functions $\{f_1, \ldots, f_n\}$ that inductively define the model. The free tree (Nourani 1995) defined by the notion of provability implied by the definition could consist of some extra Skolem functions $\{g_1, \ldots, g_m\}$, that appear at free trees. The f terms and g terms, tree congruences, and predictive diagrams then characterize partial deduction with free trees. Partial deduction can be regarded as a form of abduction reasoning, and vice versa, if differences in the underlying languages of PD and abduction are resolved (Hoppe 1992). However, the most interesting aspect of Hoppe is that predictive reasoning can be extended to perform partial deduction, if universal quantified hypotheses are used. It is when a special consistency-checking scheme, based on Hilbert's epsilon symbol (Nourani-Hoppe 1993) might be used. These extensions of PD to an abductive, nonmonotonic reasoning approach allow us to link them with the semantic approach to non-monotonic reasoning (Nourani 1991, 1995). By viewing PD from predictive diagrams we could define models for PD from predictive diagrams—thus a model theoretic formulation for PD emerges. We then define Hilbert models to handle the proof-model problems in Nourani-Hoppe (1993). To culminate model discovery we state and prove a theorem on predictive diagrams.

For planning and learning the robot's mind state can be defined by a diagram grid. The starting space applicable project was meant for autonomous robots wandering at outer space. The designs in Nourani (1991, 1993, 1995) are ways for a robot to update its mind state based on what it encountered on its path. What the robot believes can be defined on a diagram grid. The degree to which a robot believes something is on the grid. It can get strengthened or weakened as a function of what the robot learns as progress is brought on. Robot's Mind State—the array grid entries are pointing to things to remember and the degree the robot believes them. The entry 15 is an item robot believes the most.

```
 ____________________
| 3, 4, 2,3, 11,15, 0,
|
|
|
|
|____________________
```

Diagram computing on the gird.

Logically and theoretically the grid is minimally defined by the G-diagram functions. The grid model is a way to encode diagrammatic reasoning.

4.6. New Applications

Poole's theorist is a proof procedure for abductive reasoning based on Loveland (1978) MESON system. Abduction involves constructing hypotheses, where each hypothesis is a set of atomic literals Δ_i; such that when some particular theory T is augmented with Δ_i, it entails the set of goal literals G, i.e. $T \cup \Delta_i \models G$. Δ_i must be a subset of a set of ground atomic abducibles A. It addition we must ensure $T \cup \Delta_i$ is consistent under $\models$. The set of all possible hypotheses is $\Delta = \{\Delta i\}$. Abduction is minimal when the Δ_i are the minimal such sets. Abductions that are more general are defined by applying formulas that are more complex as abducibles A. First order existential formulas can form the abducibles.

According to Poole an explanation problem can be stated formally as given:

Facts $\mathcal{F}$: consistent formulas, known to be true,
Defaults Δ: possible hypotheses, that we accept as part of an explanation and
Observations $\mathcal{G}$: which are to be explained. An observation $g \in \varphi$ is explainable, if there exits a ground hypotheses $\mathcal{D} \subseteq \Delta$ such that:

1. $\mathcal{F} \cup \mathcal{D} \models g$ and
2. $\mathcal{F} \cup \mathcal{D}$ is consistent.

A symbolic computation is called an unfolding, if it substitutes according to $\vdash$ a given goal G with respect to a theory T by a conjunction of clauses.

The abstract description theorist (Goal, T, Body) $\Longleftrightarrow$ unfold (Goal, T, Body)

We have applied predictive diagrams since 1994 to partial deduction and cumulative defaults. An *predictive diagram* for a theory T is a diagram D[M], where M is a model for T, and for any formula q in M, either the function $f: q \to \{0, 1\}$ is defined, or there exists a formula p in D[M], such that $T \cup \{p\}$ proves q; or that T proves q by minimal abduction. A *generalized predictive diagram*, is an predictive diagram with D[M] defined by a specific set of functions. To state specific applications let us state the compactness theorem from standard logic and model theory.

THEOREM 4.1. *Compactness* A *formula* A *in a theory* T *is valid iff it is valid in some finitely axiomatized part of* T.

COROLLARY. *A theory T has a model iff every finitely axiomatize part of* T *has a model.*

THEOREM 4.2. *A set of first order observations $\mathcal{G}$ is explainable iff there exists a predictive diagram for the logical consequences to $\mathcal{G}$.*

PROOF. Applies the above with Godel's completeness and the compactness theorem from model theory.

($\Rightarrow$) The diagram D[M], where M is a model for T, and for any formula q in M, indicates either the function $f: q \to \{0, 1\}$ is defined, or there exists a formula p in D[M], such that $T \cup \{p\}$ proves q; or that T proves q by minimal abduction. Hence for a formula g either there exits ground hypotheses $\mathcal{D} \subseteq \Delta$ on which $f: q \to \{1, 0\}$ is defined such that $\mathcal{F} \cup \mathcal{D} \models g$; or there exists a formula p such that $T \cup \{p\}$ proves q, or T proves q my minimal abduction, hence the compactness Theorem 4.1 renders $\mathcal{F} \cup \mathcal{D}$ consistent. Thus $g \in \mathcal{G}$ is explainable.

($\Leftarrow$) When $g \in \mathcal{G}$ is explainable, there exists ground hypotheses $\mathcal{G} \subseteq \Delta$ such that:

1. $\mathcal{F} \cup \mathcal{D} \models g$
2. $\mathcal{F} \cup \mathcal{D}$ is consistent.

Therefore, either there is a function $f: g \to \{1, 0\}$ defined direct; or by appalling Godel's completeness theorem a time or two, there is a proof for g either direct or by minimal abduction. Therefore, there is an predictive diagram for the logical consequences to $\mathcal{G}$.

5. MODELS AND PROOF COMPUTATION

We had shown in Nourani-Lieberherr (1984) KR with abstract data types allows us to define automatic KR. Abstract types keep the attention of system designers on KR issues rather than programming. Even most trivial situations require masses of knowledge and data types allow us to structure knowledge. Existentially quantified diagrams for KR carry a main deficit—the Skolemized formulas are not characterized. Hilbert's epsilon symbol may be applied to solve this problem. We have presented the notion of an predictive diagram and applied it to provide a model-theoretic characterization for PD and related proof trees. An *predictive diagram* for a theory T is a diagram D[M], where M is a model for T, and for any formula q in M, either the function $f: q \rightarrow \{0, 1\}$ is defined, or there exists a formula p in D[M], such that $T \cup \{p\}$ proves q; or that T proves q by minimal abduction. Let us see what predictive diagrams do for knowledge discovery.

Predictive diagrams allow us to model-theoretically characterize incomplete KR. To key into the incomplete knowledge base we apply generalized predictive diagrams whereby specified diagram functions a search engine can select onto localized data fields. A *Generalized Predictive Diagram* is an predictive diagram with D[M] defined from a minimal set of functions. The predictive diagram could be minimally represented by the set of functions $\{f_1, \ldots, f_n\}$ that inductively define the model. The free trees defined by the notion of provability implied by the definition, could consist of some extra Skolem functions $\{g_1, \ldots, g_m\}$, that appear on the free trees. The f terms and g terms, tree congruences, and predictive diagrams then characterize partial deduction with free trees.

An example planning algorithm based on the above is that free-proof tree planning algorithm (Nourani 1994) is on a following a chapter and stated here for self-containment. We start with some Skolem functions to define Initial models. Thus we have Hilbert models for which the Skolem functions implicitly define membership to the set defining a model. For example can refer to the foundations of mathematics for the definition of epsilon models of set theory. Thus we can refer to such models in our approach as Hilbert models. How to address the issues of proof trees in connection with the formulation in Poole (1987).

In planning applications there is goal formula to be satisfied, rather than explained, from a set of premises that are facts and defaults. The applications call for an algorithm for proof abstraction, planning and EBG-Explanation-based Generalization. The computation method for goal satisfiability allows ground instances of defaults (those without variables and Skolem constants). We present the following algorithm for planning with free proof trees.

Skolemized Virtual Proof Tree Planning Algorithm

1. Skolemize F, the set of facts, forming F_{sk} (free variables are universally quantified)
2. Try to prove g, from the elements of F_{sk} and defaults DF as axioms. Let DF_0 be the instances of default used in the proof.
3. Form DF_1 by replacing free variables in DF_0 with unique constants.
4. Add DF_1 to F and try to prove an inconsistency. If a complete search for a proof fails, the goal of the plan is satisfied.

In step (3) above it might be possible to replace free variables with elements of F_{sk}, and at times might make free proof more efficient.

5.1. Discovery Computation as Plan Proof Trees

Data discovery from KR on diagrams might be viewed as satisfying a goal by getting at relevant data which instantiates a goal. The goal formula states what relevant data is sought. There are some techniques that are made use of in some recent AI literature (10) which advocate the notion of partial deduction. We propose methods that can be applied to planning with GF-diagrams such that similar effects to partial deductions can be achieved and perhaps some of the techniques can be applied to implement discovery planning. In planning with GF-diagrams that part of the plan that involves free Skolemized trees is carried along with the proof tree for a plan goal. The idea is that if the free proof tree is constructed then the plan has a model in which the goals are satisfied. The model is the initial model of the AI world for which the free Skolemized trees were constructed. Partial deductions in this approach correspond to proof trees that have free Skolemized trees in their representation. Use of nondeterminism by GF-diagrams is useful in knowledge discovery since there usually are many indeterminate formulas in the world diagram, whose actual values are inconsequential in an immediate plan of action, thus may be assigned a free Skolem function to satisfy a goal.

5.2. Generic Free Skolemized, Discovery, and Proof Abstraction

The discovery planning process at each stage can make use of GF-diagrams by taking the free interpretation, as tree-rewrite computations as in AI (1980), for example, of the possible proof trees that correspond to each goal satisfiability. There are several ways in which such planning can be achieved. The method that is proposed here is to make use of the free Skolemized proof trees in representing plans in terms of generalized Skolem functions. Planning with GF-diagrams carries along that part of the plan that involves free Skolemized trees, with the proof tree for a plan goal. The idea is that if the free proof tree is constructed then the plan has a model in which the goals are satisfied. The model is the initial model of the AI world for which the free Skolemized trees were constructed. Partial deductions in this approach correspond to proof trees that have free Skolemized trees in their representation.

Perhaps there is analogy that can bring the applications of the present formulations to the AI trends closer at hand. We could view, on the one hand, SLD resolution type proofs on ground terms, where we go from $p(0)$ to $p(f(c))$; or form $p(f(c))$ to $p((f(g(c)))$. Whereas, while doing proofs with free Skolemized trees we are facing proofs of the form $p(g(\ldots))$ proves $p(f(g(\ldots))$ and generalizations to $p(f(x))$ proves for all x, $p(f(x))$. Thus the free proofs are in some sense an abstract counterpart of the SLD proofs, in the sense of the trends that have appeared in the recent AI literature for example, EBG. These projects will be treated by our subsequent publications. Of course, the author since 1993 was also concerned with domain abstraction and program proof abstraction, while proofs were by tree rewriting. A free proof tree is a proof tree that is constructed with free Skolem functions and GFS from a GF-diagram. The idea is that if the free proof tree is constructed for a goal formula, then from the GF-diagram there is a model satisfying the goal formula satisfied. The model is the initial model of the AI world for which the free Skolemized trees were constructed.

5.3. Initial Models

To prove Godel's completeness theorem Henkin style (Henkin 1949) proceeds by constructing a model directly from the syntax of the given theory. The structure is obtained by putting terms that are provably equal into equivalence classes, then defining a free structure on the equivalence classes. The computing and reasoning enterprise require more general techniques of model construction and extension, since it has to accommodate dynamically changing world descriptions and theories. Our techniques for model building as applied to the problem of AI reasoning allows us to build and extend models through diagrams. This requires us to focus attention on the notion of generalized diagram. The minimal set of function symbols are those with which a model can be built inductively. We focus our attention on such models, since they are Initial (Goguen 1973) and computable (Nourani 1980). We apply the canonical models since they are initial (Nourani 1980) and computable (Nourani 1980, 1983).

A technical example of algebraic models defined from syntax had appeared in defining initial algebras for equational theories of data types (AI 1980) and our projects since 1978. In such direction for computing models of equational theories of computing problems are presented by a pair (Σ, E), where Σ *is a signature* (of many sorts, for a sort set S) and E a set of Σ-equations. Let $T\langle\Sigma\rangle$ be the *free tree word algebra of signature* Σ. The quotient of $T\langle\Sigma\rangle$, the word algebra of signature Σ, with respect to the Σ-*congruence* relation generated by E, will be denoted by $T\langle\Sigma, E\rangle$, or $T\langle P\rangle$ for presentation P. $T\langle P\rangle$ is the "initial" model of the presentation P. The Σ-congruence relation will be denoted by $\equiv_P$. One representation of T(P) which is nice in practice consists of an algebra of the canonical representations of the congruence classes. This is a special case of generalized standard models defined here.

In what follows $gt_1 \ldots t_n$ denotes the formal string obtained by applying the operation symbol g in Σ to an n-tuple t of arity corresponding to the signature of g. Furthermore, g_C denotes the function corresponding to the symbol g in the algebra C.

DEFINITION 5.3. Let Σ be an S-sorted signature. Then a canonical term Σ-algebra (Σ-CTA) is a Σ-algebra C such that

(1) $|C|$ is a subset of $T\langle\Sigma\rangle$ as S-indexed families
(2) $gt_1 \ldots t_n$ in C implies t is in C and; $g_C\ (t_1 \ldots t_n) = gt_1 \ldots t_n$. For constant symbols (2) must hold as well, with $g_C = g$.

THEOREM 5.3. *Let* C *be a* Σ*-algebra. Let* P =(Σ, E) *be a presentation. Then* C *is* Σ *isomorphic to* $T\langle P\rangle$, *iff*

(i) C *satisfies* E;
(ii) $g_C\ (t_1 \ldots t_n) \equiv gt_1 \ldots t_n$ *Note*: (ii) *must also hold for constants with* $g_C = g$; $\equiv$ *refers to the* Σ*-congruence generated by E.*

PROOF. see Goguen et al. (1978).

Canonical Models from models to set theory had been stated for arbitrary structures as follows.

DEFINITION 5.4. Let $[M, a]_c$ in C be defined such that M is a structure for a language $\mathscr{L}$ and each constant c in C has the interpretation a in M. The mapping $c \rightarrow a_c$ is an assignment of C in M. We say that $[M, a]_c$ in C is canonical model for a presentation P on language $\mathscr{L}$, iff the assignment $c \rightarrow a$ maps C onto M, i.e. $M = [a\colon c \text{ in } C]$.

Generic diagrams allow us to define canonical models with specific functions. The CTA's are specific canonical models. In the sections above we have shown how to define models with G-diagrams. Initial models are models definable by G-diagrams. Let us carry the initial models onto the planning foundations.

THEOREM 5.4. *For the free proof trees defined for a goal formula from the GF-diagram there is an initial model satisfying the goal formulas. It is the initial model definable by the GF-diagram.*

PROOF. In planning with GF-diagrams plan trees involving free Skolemized trees is carried along with the proof tree for a plan goal. The idea is that if the free proof tree is constructed then the plan has a model in which the goals are satisfied. There is analogy to SLD proofs. We can view on the one hand, SLD resolution type proofs on ground terms, where we go from $p(0)$ to $p(f(c))$; or form $p(f(c))$ to $p(f(g(c))$. Whereas, while doing proofs with free Skolemized trees we are facing proofs of the form $p(g(\ldots))$ proves $p(f(g(\ldots))$ and generalizations to $p(f(x))$ proves for all x, $p(f())$. Since the proof trees are either proving plan goals for formulas defined on the GF-diagram, or are computing with Skolem functions defining the GF-diagram, by Definitions 5.3, 5.4, and Theorem 5.3 the model defined by the GF-diagram applies and it is initial for the proofs.

6. DESCRIPTIVE COMPUTING

6.1. The Basics

We present what we refer to by Descriptive Computation applying generalized diagrams. We define descriptive computation to be computing with G-diagrams for the model and techniques for defining models with G-diagrams from the syntax of a logical language. G-diagrams are diagrams definable with a known function set. Thus the computing model is definable by G-diagrams with a function set. An example function set might be $\Sigma 1$ Skolem functions. The corresponding terminology (Martin 1977) in set theory refers to sets or topological structure definable in a simple way. Thus by descriptive computation we can address artificial intelligence planning and theorem proving, for example. The latter computational issues are pursued by the author in Nourani (1994, 1997b).

The logical representation for reaching the object might be infinitary only. We have seen above that the artificial intelligence problem from the robot's stand point is to acquire a decidable descriptive computation for the problem domain. The author has proved (Nourani 1994) two specific theorems for descriptive computing on diagrams. A compatibility theorem applies descriptive computing to characterize situation compatibility. Further, a computational epistemic reducibility theorem is proved by the descriptive computing techniques on infinitary languages in Chapter 6. A deterministic epistemics is defined and it is proved not reducible to known epistemics. A set is Descriptive Computable iff it is definable by a G-diagram with computable functions. Applying admissible sets we can prove that for A an admissible computable set, A is descriptive computable.

PROOF. Nourani (1997b, c).

6.2. Boolean Computing Diagrams

Applying G-diagrams to define knowledge with Skolem functions the computation the Skolemized formulas are defined and expanded to assign truth values to the atomic formulas. Boolean algebras can be defined corresponding to the formulas with Skolemized trees. A truth valuation can be defined with respect to a class of Boolean algebras $\mathfrak{R}$ as a homomorphism from a closure of formulas to an algebra in $\mathfrak{R}$.

For example we can define $\mathrm{B} = \langle B, \min, \max, -, 0, 1\rangle$ where min and max are defined by induction on depth of Skolemized trees. By the theorem below from Richter (1978) logical implication for formulas and models for classes of Boolean algebras can be reduced to only the Boolean algebra over $\{0, 1\}$.

THEOREM 6.6. *If $\mathfrak{R}$ is the class of Boolean algebras and $\mathfrak{R}'$ contains only the Boolean algebra over $\{0,1\}$ then $\models$ for $\mathfrak{R}$ is equivalent to $\models$ for $\mathfrak{R}'$.*

As we defined in Nourani (1995) a Plan is a sequence of operations in the universe that could result in terms that instantiate the truth of a specific plan goal formulas in the universe. Thus a plan can be represented by a tree. In planning with G-diagrams the plan trees involving free Skolemized trees are carried along with the proof tree for a plan goal. We call such trees *Free Proof Trees.* The idea is that if the free proof tree is constructed then the plan has a model in which the goals are satisfied. The model is the standard model definable by the generic diagram for which the free Skolemized trees were constructed. Thus we the following Free Proof Tree Sound Computing Theorem.

THEOREM 6.7. *For the free proof trees defined for a goal formula from the G-diagram there is a standard model satisfying the goal formulas. It is the model definable by the G-diagram.*

PROOF. A free proof tree is a proof tree that is constructed with free Skolem functions from a G-diagram. The idea is that if the free proof tree is constructed for a goal formula, then from the G-diagram there is a model satisfying the goal formula satisfied. The model is the standard model for which the free Skolemized trees were constructed. Thus we have transformed the model theoretic problems of computing to that of descriptive computing with G-Diagrams. What it implies for Skolemized trees and Boolean algebras is the computation by Theorem 6.8.

THEOREM 6.8. *There is a Boolean algebra $\mathfrak{R}'$ in the class $\mathfrak{R}$, $\mathfrak{R}'$ containing only $\{0,1\}$, such that for the standard model A defined by the G-diagram Theorem 5.2. $A \models$ is equivalent to $\models$ for $\mathfrak{R}'$.*

PROOF. Follows from Richter (1978).

7. CONCLUSIONS

Generalized diagrams are shown to be an encoding for a minimal efficient KR technique applied to define relevant world models and KR for data search and discovery. We have further shown how by defining predictive diagrams, partial deduction and predictive KR can be represented model-theoretically. We have also applied the techniques to proof abstraction (Nourani–Hoppe 1993) and other related problems appearing in the recent AI literature. It might delude an uninitiated that abstracting the model by G-diagrams only turns it all to formulas hence computationally difficult. On the contrary the G-diagram technique is a way to encode models with functions hence by many orders of magnitude diminish the formula complexity applied to define models. Model checking is simplified to the same degree. New areas where our papers apply (Nourani 1997a) are KR cardinality concepts (Badder et al. 1996). Proof abstraction is yet another area where KR can be keyed to KB and applied to knowledge discovery. We have defined various notions of generalized diagrams for KR applications and KB data discovery. KB completions can be applied with our techniques. Generalized diagrams are shown to be an encoding for a minimal efficient KR technique applied to define relevant world models and KR for data search and discovery. We have further shown how by defining predictive diagrams, partial deduction and abductive KR can be represented model-theoretically. We have also applied the techniques to proof abstraction and other related problems appearing in the recent AI literature.

It might delude an uninitiated that abstracting the model by G-diagrams only turns it all to formulas hence computationally difficult. On the contrary the G-diagram technique is a way to encode models with functions hence by many orders of magnitude diminish the formula complexity applied to define models. Model checking is simplified to the same degree. Initial models are presented for diagram computations. An algorithm is presented for automated theorem proving and plan computation with virtual proof trees. Model computation is facilitated with Boolean computing models.

EXERCISES

1. Apply the incomplete diagram examples with indeterminate symbols to complete a plan to design a blocks world problem solver from arbitrary incomplete diagrams for the possible blocks worlds as a robot might.
2. Where does compactness apply to prove theorem 10.2 and in what exact way?
3. Apply the algorithm in Section 5.2 to solve planning problems for the blocks world.
4. Apply Exercise 3 to plan for the geometric object morphings in Chapter 4, Section 4.2.
5. Apply predictive diagrams to design obstacle avoidance for virtual world computing, Start with the specific worlds with two robots and specified geometric object worlds as in Exercise 4.
6. Apply the unfolding algorithm with partial deduction to complete a plan goal computation. Compare your computation to a completing a G-diagram to satisfy the same goal with models.
7. Prove Theorem 6.8.

REFERENCES

1. F. Badder, M. Buchheit, and B. Hollunder, *Artificial Intelligence* 1996.
2. J. Bailey, M. Georgeff, D. B. Kemp, D. Kinny, and K. Ramamohuarao, "Active Databases and Agent Systems." A Comparison, Techncial Report. University of Melbourne, Parkville, 1995.
3. J. Barwise, "The Syntax and Sematics of Infinitary Languages," Vol. 72. Lecture Notes on Mathematics, Springer-Verlag, 1968.
4. C. C. Chang and H. J. Keisler, "Model Theory," Vol. 73. North-Holland, 1973.
5. K. Dekker and V. Lesser, "Analyzing the Need for Metalevel Communications," In Proc. of the Twelfth International Workshop on Distributed AI, Hidden Valley, Pennsylvania, 1993.
6. Fikes, R. E. and N. J. Nilsson, *Artificial Intelligence* 2, 189 (1971).
7. J. Galsgow, N. Hari Narayanan, and B. Chandrasekaran, Eds., "Diagramatic Reasoning: Cognitive and Computational Perspectives" p. 807. AAAI Press, 1995.
8. M. Genesereth and N. J. Nilsson, "Logical Foundations of Artificial Intelligence," Morgan-Kaufmann, 1987.
9. J. A. Goguen, J. W. Thatcher, E. G. Wagner, and J. B. Wright, "A Junction Between Computer Science and Category Theory." IBM Research Report RC 4526, 1973; J. Barwise, "Handbook of Mathematical Logic." 2nd ed. North-Holland, 1978.
10. J. A. Goguen, J. W. Thatcher, E. G. Wagner, and J. B. Wright "An Algebraic Approach to Abstract Data Type Specifications and Their Correctness of Implementations" (R. Yeh, Ed.), Vol. 4. Current Trends in Programming Methodology, Data Structuring Techniques, Prentice Hall, 1978.
11. T. Hoppe, "On The Relationship Between Partial Deduction and Abductive Reasoning," pp. 154–158. Proc. 10th ECAI Conference, John Wiley and Sons, Vienna, Austria, 1992.
12. D. W. Loveland, "Automated Theorem Proving: A Logical Basis." North Holland, 1978.

13. D. A. Martin, "Descriptive Set Theory: Projective Sets," pp. 783–815. Handbook Mathematical Logic, 1977.
14. C. F. Nourani, "A Model-theoretic Approach to Specification, Extension, and Implementation of Data Type," Proc. International Symposium on Programming. Springer-Verlag, Berlin, 1980.
15. C. F. Nourani, *J. Association Computing Machinery* 30, 343 (1983).
16. C. F. Nourani, "Equational Intensity, Initial Models, and AI Reasoning," Technical Report: A Conceptual Overview, in Proc. Sixth European Conference in Artificial Intelligence. Pisa, North-Holland, Italy, 1984.
17. C. F. Nourani, "Diagrams, Possible Worlds, and the Problem of Reasoning in Artificial Intelligence." Logic Colloquium, Padova, Italy, 1988.
18. C. F. Nourani, "Planning and Plausible Reasoning in AI," pp. 150–157. Proc. Scandinavian Conference in AI. IOS Press, Roskilde, Denmark, 1991.
19. C. F. Nourani, "How Could There be Models for Nothing and Proofs for Free." IDLP, Darmstadt, Germany, 1993.
20. C. F. Nourani, "Towards Computational Epistemology—A Forward." Summer Logic Colloquium, Clairemont-Ferrand, France, 1993.
21. C. F. Nourani, "Computation on Diagrams, and Free Skolemization." Summer Logic Colloquium, England, 1993.
22. C. F. Nourani, "Towards Computational Epistemology—A Forward," Proc. Summer Logic Colloquium, Clermont-Ferrand, France, 1994.
23. C. F. Nourani, "Free Proof Trees and Model-Theoretic Planning." Automated Reasoning, AISB, England, 1995.
24. C. F. Nourani "KR, Diagrams For Models, and Computation." ASL, Toronto, 1996.
25. C. F. Nourani, "Descriptive Computing." Summer Logic Collqouium San Sebstian, AMS, Memphis, Spain, 1997a.
26. C. F. Nourani, "Descriptive Computing-The Preliminary Definition," Summer Logic Colloquium. AMS, Memphis, San Sebstian Spain, 1997b.
27. C. F. Nourani, "Functorial Admissible Models," ASL, MIT, Cambridge, 1997c.
28. C. F. Nourani, "Computability, KR and Reducibility," ASL, Toronto, 1998a.
29. C. F. Nourani, "Syntax Trees, Intensional Models, and Modal Diagrams for Natural Language Models." Uppsala Logic Colloquium, Uppsala University, Sweden, 1998b.
30. C. F. Nourani and T. Hoppe, "GF Diagrams, and Free Proof Trees." Technical Report, 1993; I. Mosetic and C. Holsbaur, "Extending Explanation Based Generalization by Abstraction Operators," Vol. 482. Machine Learning EWSL-91, Springer-Verlag, LNAI, 1991.
31. C. F. Nourani and T. Hoppe, "G-Diagrams for Models, Free Proof Trees, and Predictive Automated Reasoning." Automated Reasoning-AISB, England, 1995.
32. C. F. Nourani and K. J. Lieberherr, "Data Types, Direct Implementations and KR." 19th HICSS, KR Track, Hawaii, 1986.
33. D. Poole, R. Goebel, and R. Aleliunas, "Theoreist: A Logical Reasoning System For Defaults and Diagnosis" (N. Cercone and G. McCalla, Eds.). The knowledge Frontier, Essays in the Representation of Knowledge. Springer-Verlag, 1987.
34. M. M. Richter, "Logikkalkule," Teubner Studeinbucher, Tubner, Leibzig, 1978.
35. Special Issue on Non-monotonic Logic, *Artificial Intelligence* 13, (1980).

CHAPTER 11

Business Intelligence, Content Exploration, and Intelligent Interfaces

CONTENTS

ISBN: 1-58883-037-3/$35.00

Intelligent Multimedia Computing Science
By Cyrus F. Nourani
Pages: 137–152

1. INTRODUCTION

Business intelligence interfaces might apply automated learning and discovery—often called data mining, machine learning, or advanced data analysis with new w-interface relevance. There are obvious financial and organizational memory applications applied at times in our projects. E-Commerce, e-business, trust, trustworthiness, usability, human-computer interaction, cognitive ergonomics, user interface design, ease of use, interaction design, and online marketing, are the business user modeling issues the paper addresses. Financial companies have begun to analyze their customers' behavior in order to maximize the effectiveness of marketing efforts. There are routine applications to data discovery techniques with intelligence databases. Management process controls at times calls on warehoused data and relies on organizational memory to reach decision. Recent research has led to progress, both in the type methods that are available, and in the understanding of their characteristics. Intelligent business interfaces are designed with intelligent multi-tiers applying agents and intelligent business objects with applications to intelligent WWW interfaces. Basic intelligent content management with multi-tier designs for interfaces are presented. The field of automated learning and discovery has obvious financial and organizational memory applications. There are basic applications to data discovery techniques with intelligence multimedia databases. The computing model is based on a novel competitive learning with agent multiplayer game tree planning.

Knowledge management—KM and organizational memory (OM) are areas where model discovery with active intelligent databases applying predictive logic presented here applying intelligent multimedia databases. Knowledge management (KM) is one of the key progress factors in organizations. In an organization, expertise may relate to problem solving expertise in functional disciplines, experiences of human resources, and project experiences in terms of project management issues, design technical issues and lessons learned. "Corporate memory" or "organizational memory" is regarded as the central prerequisite for IT supports of knowledge management. It is the means for knowledge conservation, distribution, and reuse. Identification and analysis of a company's knowledge-intensive work processes (e.g., product design or strategic planning, utilizing KM and OM are datadiscovery and datawarehousing intensive operations.

2. THE VISUAL DYNAMICS

2.1. Why Multimedia Databases?

Defining compatibility and visual effects relations, allows objects to be selected and applied to design and compare customized views. Multimedia programming is combined with visual multiagent objects to define specific visual compatibility for customized active databases. Active databases deploy certain computing which lend themselves naturally to the inelegant multimedia principles Nourani (1999c). The concept of an active objects are embedded by intelligent objects and "events" are embedded by the computing defined by IM as a basic principle an embedded by intelligent trees, intelligent objects, and hybrid pictures. The characteristics of an Active DMBS, or ADMBS, see for example Bailey (1995), supports definition and management of ECA-rules, e.g. event, condition, and action. Hence an ADMBS must have means to define ECA's. An ADBMS must support rule management and rule-base updates. It must carry out actions and evaluate conditions. An ADBMS must represent information in ECA-rules in terms of its data models. The IM computing paradigm provides a basis for designing multimedia ADBMS's. The IM computing paradigm allows the design for multimedia ADBMS Nourani (2000) to apply agent computing to ADBMS, to base an ADBMS on multimedia agent computing, and to carryout metalevel reasoning and KB with multimedia intelligent objects.

3. THE MODELS

3.1. Intelligent Models

A computational logic for intelligent languages is presented in brief with a soundness and completeness theorem in Nourani (1999a) from Nourani (1994) and applied to intelligent business object computing. A brief overview to context abstraction shows how context free and context sensitive properties might be defined with agents. Agents are in the sense defined in Nourani (1996b) and the AI foundations Genesereth (1987). A set of function symbols in the language, referred to by agent function set, is the set of function symbols that are modeled in the computing world by AI agents. The idea is to characterize computation with abstract syntax trees without grammar specifics. As an example, suppose you are told there is an academic department with a faculty member which is superman, and two faculty members which are Swedish speaking, and three which do not talk to anybody not in their expertise area. Without telling you anything else about what they do, abstract syntax properties can be defined. When the signature's specific agent functions are defined, it implies the signature has defined message paths for them. From the signature intelligent models are defined for abstract syntax trees. The computation is expressed by an abstract language that is capable of specifying modules, agents, and their communications. The implementing agents, their corresponding objects, and their message passing actions can also be presented by the two-level abstract syntax. The agents are represented by function names that appear on the free syntax trees of implementing trees. The trees defined by the present approach have function names corresponding to computing agents. The computing agent functions have a specified module defining their functionality. A signature defines the language tree compositionality degree and defines the abstract syntax. The definitions prompted a computational linguistics and model theory for intelligent languages. Models for the languages are defined by our techniques in Nourani (1996b, 1999b).

3.2. Relevant KR and Visual Business Objects

Knowledge representation has two significant roles: to define a model for the AI world, and to provide a basis for reasoning techniques to get at implicit knowledge. An ordinary diagram is the set of atomic and negated atomic sentences that are true in a model. Generalized diagrams are diagrams definable by a minimal set of functions such that everything else in the model's closure can be inferred, by a minimal set of terms defining the model. Thus providing a minimal characterization of models, and a minimal set of atomic sentences on which all other atomic sentences depend. We want to solve real world problems in AI. Obviously, for automating problem solving, we need to represent the real world. Since we cannot represent all aspects of a real world problem, we need to restrict the representation to only the relevant aspects of the real world we are interested in. Let us call this subset of relevant real world aspects the relevant world for a problem. AI approaches to problem solving represent the knowledge usually in some kind of first-order language, consisting of at least constants, functions and predicate symbols. Our primary focus will be the relations amongst KR, AI worlds, and the computability of models. Truth is a notion that can have dynamic properties. To keep the models which need to be considered small and to keep a problem tractable, we have to get a grip on a minimal set of functions to define computable models with. The selector functions are applied to create compound business objects.

3.3. Learning from Multimedia Data

Many data sets contain more than just a single type of data. The existing algorithms can usually only cope with a single type data. How can we design methods that can take on multimedia data from multiple modalities? Are we to apply separate learning algorithms to each data modality, and combine their results, or are there to be algorithms that can handle multimedia multiple modalities on a feature-level. Learning casual relationships amongst visual stored data is another important area, which can benefit from our project. Most existing learning algorithms detect only correlations, but are unable to model causality and hence fail to predict the effect of external controls. Visualization and interactive discovery data mining is a process which involves automated data analysis and control decisions by an expert of the domain. For example, patterns in many large-scale business system databases might be discovered interactively, by a human expert looking at the data, as it is done with medical data. Data visualization is specifically difficult when data is high-dimensional, specifically when it involves non-numerical data such as text. The projects might be a basis for designing interactive business tools.

3.4. Computable World Models

The techniques in Nourani (1991, 1995, 1999b) for model building as applied to the problem of AI reasoning allows us to build and extend models by diagrams. This requires us to define the notion of generalized or generic diagram. The G-diagrams are used to build models with a minimal family of generalized Skolem functions. The minimal sets of function symbols are those with which a model can be built inductively. We focus our attention on such models, since they are computable (Bailey et al. 1995, Nourani 1991). The G-diagram methods applied and further developed here, allows us to formulate AI world descriptions, theories, and models in a minimal computable manner. It further allows us to view the world from only the relevant functions. Thus models and proofs for the specified problems can be characterized by models computable by a set of functions. The G-diagram functions can define IM objects and be applied with the Morph Gentzen logic.

4. HETEROGENEOUS COMPUTING AND MODELS

4.1. Heterogeneous Computing

In Nourani (1998) we present new techniques for design by software agents and new concepts entitled abstract intelligent implementation of AI systems (AII). Objects, message passing actions, and implementing agents are defined by syntactic constructs, with agents appearing as functions. The techniques have been applied to design intelligent business objects in Nourani (1998). AII techniques have been applied to heterogeneous KB design and implementation. The application areas include support for highly responsive planning. AII techniques are due to be an area of crucial importance as they are applied gradually to the real problems. The applied fields are, for example, intelligent business systems, aerospace, AI for robots, and multimedia.

4.2. KR, KB, and Visual Model Discovery

Model diagrams allow us to characterize incomplete KR. To key into the incomplete knowledge base we apply generalized predictive diagrams whereby specified diagram functions a search engine can select onto localized data fields. The predictive model diagrams Nourani (1995, 1999a) could be minimally represented by the set of functions $\{f_1, \ldots, f_n\}$ that inductively define the model. Data discovery from KR on diagrams might be viewed as satisfying a goal by getting at relevant data which instantiates a goal. The goal formula states what relevant data is sought. We propose methods that can be applied to planning Nourani (1991) with diagrams to implement discovery planning. In planning with G-diagrams that part of the plan that involves free Skolemized trees is carried along with the proof tree for a plan goal. Computing with diagram functions allows us to key to active visual databases with agents.

4.3. Prediction and Discovery

Minimal prediction is an artificial intelligence technique defined since the author's model-theoretic planning project. It is a cumulative nonmonotonic approximation attained with completing model diagrams on what might be true in a model or knowledge base. A predictive diagram for a theory T is a diagram D(M), where M is a model for T, and for any formula q in M, either the function $f\colon q \rightarrow \{0, 1\}$ is defined, or there exists a formula p in D(M), such that $T \cup \{p\}$ proves q; or that T proves q by minimal prediction. A generalized predictive diagram, is a predictive diagram with D(M) defined from a minimal set of functions. The predictive diagram could be minimally represented by a set of functions $\{f_1, \ldots, f_n\}$ that inductively define the model. The free trees we had defined by the notion of provability implied by the definition, could consist of some extra Skolem functions $\{g_1, \ldots, g_l\}$ that appear at free trees. The f terms and g terms, tree congruences, and predictive diagrams then characterize partial deduction with free trees. The predictive diagrams are applied to discover models to the intelligent game trees. Prediction is applied to plan goal satisfiablity, and can be applied with plausabilities Nourani (1991), probabilities, and fuzzy logic to obtain, for example, confidence intervals. The author has applied minimal prediction to simply encode knowledge with model diagrams to carry on automated deduction. Modeling with virtual tree planning Nourani (2002b) is applied where uncertainty, including effector and sensor uncertainty, are relegated to agents, where competitive learning on game trees determines a confidence interval. The incomplete knowledge modelling is treated with KR on predictive model diagrams. Model discovery at KB's are with specific techniques defined for trees. Model diagrams allow us to model-theoretically characterize incomplete KR. To key into the incomplete knowledge base we apply generalized predictive diagrams whereby specified diagram functions a search engine can select onto localized data fields. The predictive model diagrams could be minimally represented by the set of functions $\{f_1, \ldots, f_n\}$ that inductively define the model. Data discovery from KR on diagrams might be viewed as satisfying a goal by getting at relevant data which instantiates a goal. The goal formula states what relevant data is sought. We propose methods that can be applied to planning Nourani (1991) with diagrams to implement discovery planning. In planning with G-diagrams that part of the plan that involves free Skolemized trees is carried along with the proof tree for a plan goal. Computing with diagram functions allows us to key to active visual databases.

4.4. TAIM[1] and Discovery from Warehoused Data

The morphing active intelligent database designs outline in the paper is abbreviated as TAIM hereforth. Data discovery from KR on diagrams might be viewed as satisfying a goal by getting at relevant data which instantiates a goal. The goal formula states what relevant data is sought. We have presented planning techniques, which can be applied to implement discovery planning. In planning with G-diagrams that part of the plan that involves free Skolemized trees is carried along with the proof tree for a plan goal. The idea is that if the free proof tree is constructed then the plan has a model in which the goals are satisfied. The model is the initial model of the AI world for which the free Skolemized trees were constructed. Partial deductions in this approach correspond to proof trees that have free Skolemized trees in their representation. While doing proofs with free Skolemized trees we are facing proofs of the form $p(g(\ldots))$ proves $p(f(g(\ldots)))$ and generalizations to $p(f(x))$ proves for all x, $p(f(x))$. Thus the free proofs are in some sense an abstract counterpart of the SLD. Let us see what predictive diagrams do for knowledge discovery knowledge management. Diagrams allow us to model-theoretically characterize incomplete KR. To key into the incomplete knowledge base. The following figure depicts selector functions F_i from an abstract view grid interfaced via an inference engine to a knowledge base and in turn onto a database.

Generalized predictive diagrams are defined, whereby specified diagram functions and search engine can select onto localized data fields. A generalized predictive diagram, is a predictive diagram where D(M) is defined from a minimal set of functions. The predictive diagram could be minimally represented by a set of functions $\{f_1, \ldots, f_n\}$ that inductively define the model. The functions are keyed onto the inference and knowledge base to select via the areas keyed to, designated as Si's in Figure 1 and data is retrieved Nourani (1999e). Visual object views to active databases might be designed with the above. The trees defined by the notion of provability implied by the definition might consist of some extra Skolem

[1] TAIM is the name for the Active Intelligent Database design and is the technical property of the author's ventures. Applications must be with persmission and venture partnerships.

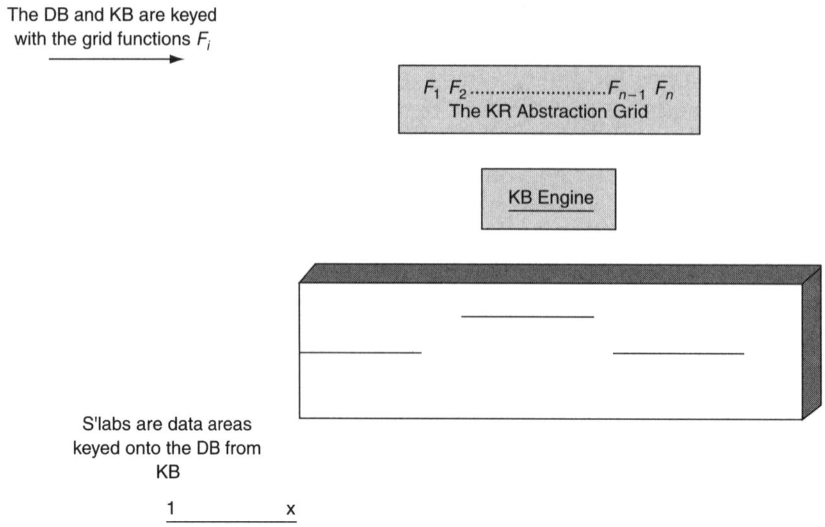

Figure 1. Keyed KR, inference, and visual discovery.

functions $\{g_1, \ldots, g_n\}$, that appear at free trees. The f terms and g terms, tree congruences, and predictive diagrams then characterize deduction with virtual trees Nourani (2002c) as intelligent predictive interfaces. Data discovery from KR on diagrams might be viewed as satisfying a goal by getting at relevant data which instantiates a goal. The goal formula states what relevant data is sought. We have presented planning techniques, which can be applied to implement discovery planning. In planning with G-diagrams that part of the plan that involves free Skolemized trees is carried along with the proof tree for a plan goal.

The idea is that if the free proof tree is constructed then the plan has a model in which the goals are satisfied. The model is the initial model of the AI world for which the free Skolemized trees were constructed. Partial deductions in this approach correspond to proof trees that have free Skolemized trees in their representation. While doing proofs with free Skolemized trees we are facing proofs of the form $p(g(\ldots))$ proves $p(f(g(\ldots)))$ and generalizations to $p(f(x))$ proves for all x, $p(f(x))$. Thus the free proofs are in some sense an abstract counterpart of the SLD. Practical AI goal satisfaction. The predictive diagram could be minimally represented by a set of functions $\{f_1, \ldots, f_n\}$ that inductively define the model. The free trees we had defined by the notion of provability implied by the definition, could consist of some extra Skolem functions $\{g_1, \ldots, g_l\}$ that appear at free trees. The f terms and g terms, tree congruences, and predictive diagrams then characterize partial deduction with free trees.

4.5. KR and Data Warehousing with Keyed Functions

Let us see what predictive diagrams do for knowledge discovery knowledge management. The preceding chapters indicate how diagrams allow us to model-theoretically characterize incomplete KR. The following figure depicts selector functions F_i from an abstract view grid interfaced via an inference engine to a knowledge-base and in turn onto a database. Practical AI systems are designed by modelling AI with facts, rules, goals, strategies, knowledge bases. Patterns, schemas, AI frames and viewpoints are the micro to aggregate glimpses onto the database and knowledge bases were masses of data and their relationships-representations, respectively, are stored. Schemas and frames are what might be defined with objects, the object classes, the object class inheritances user-defined inheritance relations, and specific restrictions on the object, class, or frame slot types and behaviors. From Nourani (2000) scheme might be intelligent forecasting.

IS-A stock forecasting technique
Portfolios stock, bonds, corporate assets
Member management science techniques

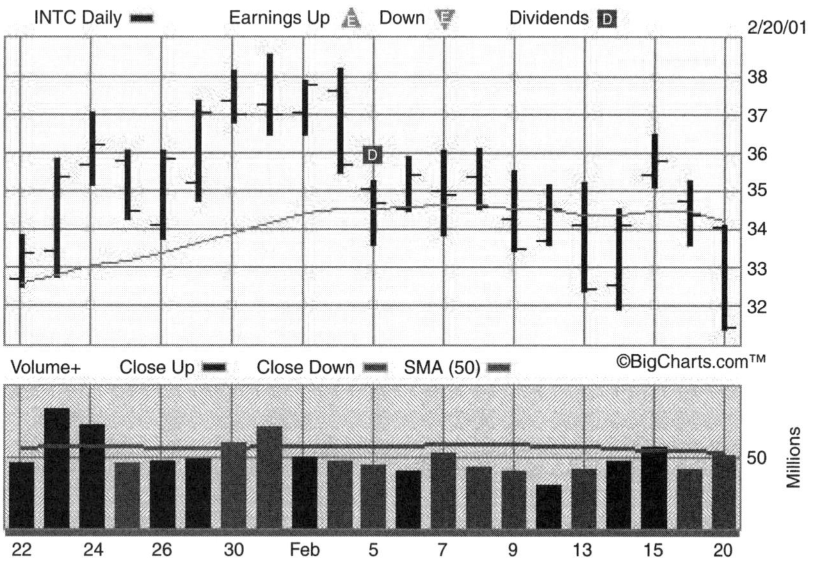

Figure 2. Stock trade interfaces.

Schemas allow brief descriptions on object surface properties with which high level inference and reasoning with incomplete knowledge can be carried out applying facts and the defined relationships amongst objects. Relationships: Visual objects have mutual agent visual message correspondence. Looking for patterns is a way some practical AI is a carried on with to recognize important features, situations, and applicable rules. From the proofs standpoint patterns are analogies to features as being leaves on computing trees. Forward chaining is a goal satisfaction technique, where inference rules are activated by data patterns, to sequentially get to a goal by apply the inference rules. The current pertinent rules are available at an agenda store.

The carried out rules modify the database. Backward chaining is an alternative based on opportunistic response to changing information. It starts with the goal and looks for available premises that might be satisfied to have gotten there. Goals are objects for which there is automatic goal generation of missing data at the goal by recursion backward chaining on the missing objects as sub-goals. Data unavailability implies search for new goal discovery. Goal directed planning is carried out while planning with diagrams. The part of the plan that involves free Skolemized trees is carried along with the proof tree for a plan goal. If the free proof tree is constructed then the plan has an initial model in which the goals are satisfied. IM's basis for forecasting is put forth at preliminary stages in Nourani (1999a).

4.6. The Stock Traders Interface Model

The Morph-Gentzen logic with predictive model diagrams has been applied as a basis for intelligent forecasting Nourani (1999a, 2000). The are graphics sequents for predicting the quarter earnings from the second and third combined with a market condition graph. The way a market condition graph is designed is a propriety issue. It is obtained by Morph Gentzen sequents from known stock market parameters. Data discovery from KR on diagrams might be viewed as satisfying a goal by getting at relevant data which instantiates a goal. The goal formula states what relevant data is sought. The basis for forecasting is put forth at preliminary stages in the author's publications since 1998. The idea is to apply Morph-Gentzen logic as a basis for intelligent multimedia forecasting. The figure indicates a graphics sequent for predicting the fourth quarter earnings from the second and third combined with a market condition graph. The way a market condition graph is designed is a propriety issue. It is obtained by Morph Gentzen sequents from known stock market parameters. The above are example stock trading interfaces.

4.7. The Visual DB Asset Management Model

The fast track to real media content management is a media asset management environment that allows vendors and customers media applications to interoperate seamlessly within the IBM DB2 digital library infrastructure. Asset management is an important area to the efficient intelligent multimedia computing. Cost-effective management of digital content is expected to be another frequent topic. IBM's grand central media "Grand Central Media" provides media asset management that enables a company to manage, secure, share, locate and reuse digital media files. The IBM solution also allows media and entertainment executives to import and index digital content and combine search requests into one results list that can be graphically browsed and shared with others world-wide. IBM grand central media is part of the company's suite of digital library applications that are being used by leading film studios, advertising firms, and broadcast and stock footage houses. The Web-based solution supports the synchronization of databases at different physical locations or on wide-area networks. The academia for example, the University of California, Berkeley's BMRC has designed a video on demand system with asset management.

4.8. Multimedia Middleware DB and Wrappers Interfaces

Large-scale multimedia information systems today rely on large collections of data stored in diverse systems with differing capabilities. Many of the "mission-critical" tasks these customers do depend on integrating data from several sources. Often the data sources to be integrated include a broad range of sources, such as CAD/CAM systems, text search engines, molecular structure databases, or customer-specific repositories. Garlic (1995) offers the ability to interrelate data from multiple sources with a broad range of querying capabilities, in a single, cross-source query. Garlic is an IBM prototype, which allows new sources to be easily added to an existing installation. Garlic offers the ability to interrelate data from multiple sources with a broad range of querying capabilities, in a single, cross-source query. A significant focus of the project is the provision of support for data sources that provide type-specific indexing and query capabilities. Garlic's "wrapper architecture," see e.g., Wiederhold (1992) and the applications at Nourani (1996a) encapsulate data sources, allowing new sources to be added quickly, and accommodating a wide variety of sources, with a broad range of traditional and nontraditional query processing capabilities. Wrappers model legacy data as objects, participate in query planning, and provide standard interfaces for method invocation and query execution. The knowledge base is keyed with relevant selector functions. There are worldwide web retrieval designs, which have to apply direct links. The hierarchical abstract structures are being applied to the knowledge base and data structures to accomplish the tasks called for.

4.9. Content Management System with TAIM

To design interfaces with systems we have to provide content management interfaces.

The intelligent interface component and the IVBM might be designed applying the three-tier model as depicted as the basis.

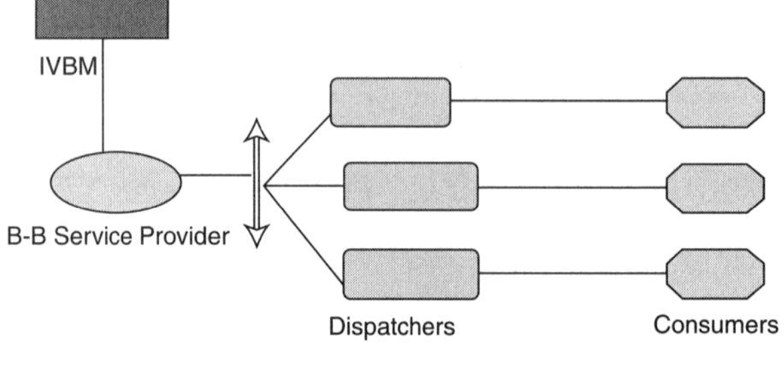

Figure 3. B-B network.

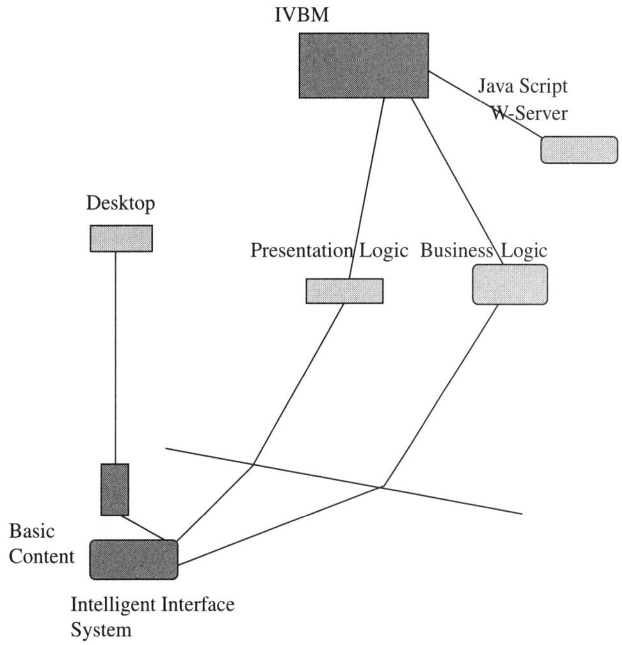

Figure 4. Intelligent B2-B interface design.
Note. The design is the author's venture properties. Application must be with permission and partnerships.

4.10. Tiers and Remote Invocation Basics

The stub implements an interface with the same business methods as the object itself, but the stubs methods do not contain business logic. The data tier consists of one or more databases and may contain data-related logic in the form of stored procedures.

5. ENTERPRISE RESOURCE PLANNING

A new basis for ERP with multiplayer games with the new computing paradigms is essential. New design techniques are presented with applications introducing intelligent business objects. The development of intelligent business objects with agents to reiterate new generations of business objects in the multimedia environment is new progress. The project designs financial software applying knowledge bases to specify the multiagent design. Practical applications with intelligent multimedia and specific business object design techniques are put forth for intelligent business objects. New direction for forecasting and business planning (Nourani 1998, 1999a) is put forth applying Morph Gentzen. We have further showed how computable AI world models for KB and knowledge management are designed. KR with G-diagrams for models is applied for keying to knowledge bases. Model discovery at KB's are

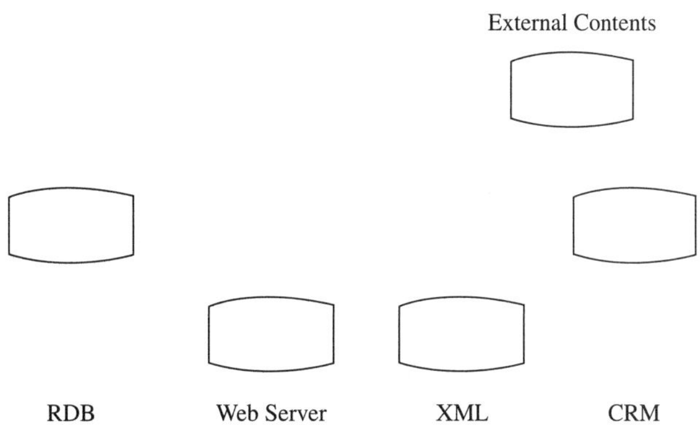

Figure 5. The content store.

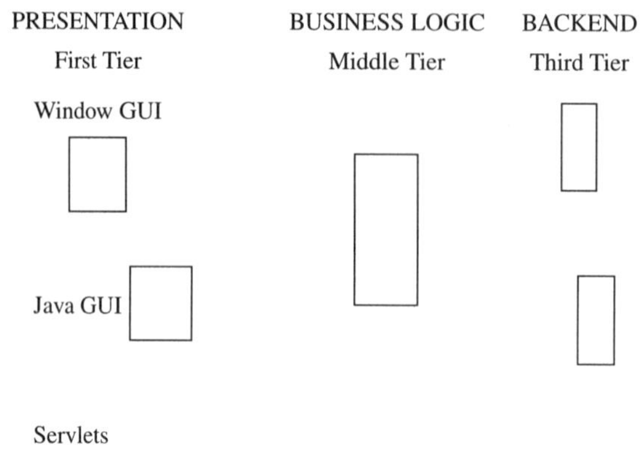

Figure 6. Tiers.

with specific techniques defined for trees. New applications to business with intelligent object languages are presented in brief. Intelligent multimedia and the new Morph Gentzen logic from the preceding chapters are applied as a preliminary basis for to forecasting. We present a brief overview to new KR techniques applied with localized keying to models G-diagram for business models and as a minimal efficient computable way for knowledge management. The corporations have to specify how to conduct business before they implement an ERP system. The best way to do that is by defining how decisions are made in terms of planning. ERP and EM are to be applied to elect supply chain policies which can in part specify how the business is to operate. Applying resource planning processes business planning models could be developed. Apply ERP to the planning process, we might develop tactical planning models that plan critical resources up to sales and delivery. Planning and scheduling require definition of their respective policies and processes; and the analysis of supply chain parameters. EM can be applied to determine the scope of the project to fit within the resources available (including budget) and the time required and to assign stake responsibility. ERP has to select the operational planning teams that will manage the process. Amongst the tasks are feasible production schedules, mapping the operational planning process to develop scheduling models for production and plan inventories. Develop plan for building a data warehouse, KM and interfaces to legacy systems and ERP. Determine the order that these parts can be implemented. ERP applies implementation teams consisting of business representatives, functional representatives, IT personnel and system knowledgeable people. The representative to be the team leader. The planning process and ERP implement schedules and raise critical success factor problems to appropriate senior management. Conducting gap and risk analysis between functionality that will be implemented and system capability. ERP and EM can be applied to determine flexibility factors to modify implementation plans as specific functionality appears more critical.

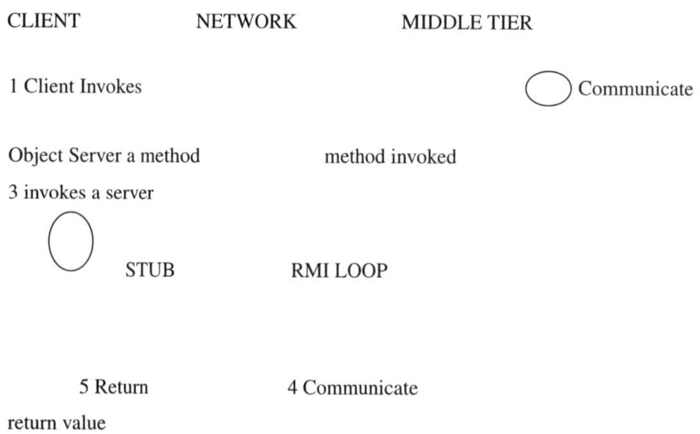

Figure 7. The RMI loop.

5.1. Infinite Multiplayer Game Trees and ERP

Games play an important role as a basis to economic theories and can be applied to ERP as a basis to EM prototypes. Intelligent tree computing theories the author defined since 1994 can be applied to present precise strategies and prove theorems on multiplayer games. Game tree degree with respect to models is defined and applied to prove soundness and completeness. The game is viewed as a multiplayer game with only perfect information between agent pairs. Upper bounds on determined games are presented. The author had presented a chess playing basis in 1997 to a computing conference. For each chess piece a designating agent is defined. The player P makes its moves based on the board B it views. P,B might view chess as if the pieces on the board had come alive and were autonomous agents carrying out two-person games as in Alice in Wonderland. Game moves are individual tree operations. The specifics are presented in the mathematical foundations chapters and the following section.

5.2. Intelligent AND/OR Trees and Splitting Agents

AND/OR trees Nilsson (1969) are game trees defined to solve a game from a player's stand point.

Formally a node problem is said to be solved if one of the following conditions hold.

1. The node is the set of terminal nodes (primitive problem—the node has no successor).
2. The node has AND nodes as successors and the successors are solved.
3. The node has OR nodes as successors and any one of the successors is solved.

A solution to the original problem is given by the subgraph of AND/OR graph sufficient to show that the node is solved. A program which can play a theoretically perfect game would have task like searching and AND/OR tree for a solution to a one person problem to a two-person game. An intelligent AND/OR tree is and AND/OR tree where the tree branches are intelligent trees.

The branches compute a Boolean function via agents. The Boolean function is what might satisfy a goal formula on the tree. An intelligent AND/OR tree is solved iff the corresponding Boolean functions solve the AND/OR trees named by intelligent functions on the trees. Thus node m might be $f(a_1, a_2, a_3)$ & $g(b_1, b_2)$, where f and g are Boolean functions of three and two variables, respectively, and a_i's and b_i's are Boolean valued agents satisfying goal formulas for f and g.

The chess game trees can be defined by agent augmenting AND/OR trees (Nilsson 1969). For the intelligent game trees and the problem solving techniques defined, the same model can be applied to the game trees in the sense of two person games and to the state space from the single agent view. The two person game tree is obtained from the intelligent tree model, as is the state space tree for agents. To obtain the two-person game tree the cross-board-coboard agent computation is depicted on a tree. Whereas the state-space trees for each agent is determined by the computation sequence on its side of the board-coboard. Thus a tree node m might be $f(a_1, a_2, a_3)$ & $g(b_1, b_2)$, where f and g are Boolean functions of three and two variables, respectively, and a_i's and b_i's are Boolean valued agents satisfying goal formulas for f and g.

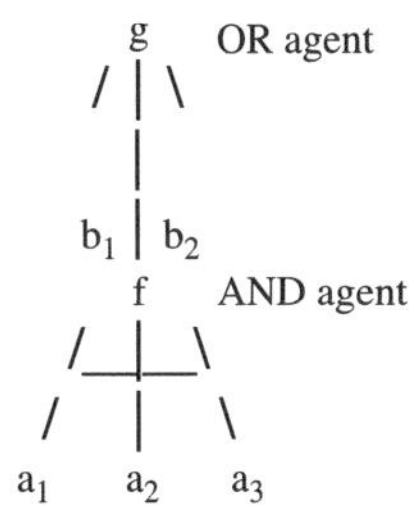

Figure 8. Agent AND/OR game trees.

A tree game degree is the game state a tree is at with respect to a model truth assignment. Game tree degrees are applied from Nourani (1999f) to decision trees and ERP models (Nourani 2002a).

6. BUSINESS INTELLIGENCE, KM, AND CONTENT RETRIEVAL

The following sections present what the current application areas are and how the above design can be applied or compared to what is addressed in multimedia knowledge management. Compare to your MDM techniques. The following section address the important areas and present the difficult challenges faced to reach at least the preliminary stages where the areas might be accommodated. It becomes apparent to the reader how the multiagent multimedia techniques presented in the book address and accommodate the challenges. Some specific areas are stated as exercises to the chapter.

6.1. Business Intelligence and Data Warehousing

To compare to example systems the SQL Server Yukon Beta 1 deploys relational data warehousing. The business intelligence platform for SQL Server "Yukon" Beta 1 based on an overview written by Joy Mundy. The Microsoft SQL Server "Yukon" provides the features, tools, and functionality to basic or innovative analytics. BI development "Workbench" is the first integrated development environment designed for the business intelligence system developer. Built on visual studio. Data transformation services can perform complex data integration, transformation, and synthesis at high speed for very large data volumes. A seamless view of data from heterogeneous sources are generated from the analytics. While the SQL Server "Yukon" blurs the lines between relational and multidimensional databases in the above intelligent interface model fills the RDB box, and the design yet has to deploy the intelligent content IVBM TAIM. You can store data in the relational database, in the multidimensional database, or use the new proactive cache feature to get the best of both worlds.

Microsoft's data mining is applied to analyze with the addition of important new algorithms, including association rules, time series, regression trees, sequence clustering, neural nets, and Naïve bayes. Important new analytical capabilities have been added to Analysis Services cubes as well: Key performance indicator framework, MDX scripts, and other built-in advanced business analytics. The business intelligence platform reaches the business user who needs to consume the analysis. Reports can be personalized and delivered in a variety of formats, with a range of interactivity and printing options. Complex analysis can reach a broad audience through the distribution of reports as a data source for downstream business intelligence. Microsoft's and partners' *ad hoc* query and analysis tools will continue to be a popular choice for accessing data in analysis services and relational databases.

The key goal of the SQL Server "Yukon" business intelligence components is to support the development and use of business intelligence in enterprises of all sizes, and to all employees—not just management and analysts, but also operational and external constituents, can be applied to the ERP areas above. The latter areas are what are important to the ERP designs on the preceding sections. However, the content management and retrieval components for business intelligence is what can complement the TAIM intelligent multimedia interfaces presented in Section 5.10 and does not include the intelligent processing presented there.

6.2. IM Knowledge Management

The new diverse media processing areas are starting to be integrated to form systems for intelligent multimedia knowledge management. Several systems and projects are reported in Jones (2003). The knowledge management activities, for example extracting information from various media such as text, speech and video, and integrating it to form alternative enriched media streams.

The applications include business, e.g. by automatically providing additional relevant information during meetings, knowledge discovery and integration from different sources, education and entertainment. The component technologies required by these systems include some or all of information retrieval, information extraction, speech recognition, video and image analysis, summarization, agent technologies, multilingual systems, database systems and information visualization. Intelligent knowledge management seeks to further integrate technologies to actually mine knowledge from documents and automatically make use of it. Extraction of information from multimedia data requires the development and integration of robust event detection and named entity recognition.

A variety of technologies have been developed to index and manage information contained within unstructured documents and information sources. Although often originally developed for textual data many of these information management tools have been applied successfully to imperfectly indexed multimedia data. It is the author's view that the indexing techniques have to be augmented with the technologies presented in the chapters here, for example, TAIM's technologies. Multimedia data presents new challenges to develop more sophisticated tools to extract, manage and combine knowledge from multiple sources. The prevalent techniques were primarily designed for textual data and vocabularies. Multimedia data lacks an obvious descriptive vocabulary, thus it is unclear how to index the content. The integration of existing techniques while a preliminary prerequisite to managing data in part, is not the solution to multimedia content problems and multimedia knowledge management. Jones (2003) states open questions remain based on technology limitations. There are techniques ranging form content and context abstraction to the various multimedia KR bases presented in the preceding chapters. Content indexing is not the ultimate solution to the content problem. The TAIM project is towards new techniques to address the challenging problems. Digital archives are another resource for automatic access to multimedia information. However, archives most commonly contain textual material, but increasingly content is derived from other media or multimedia sources including audio and visual content. Archives typically consist either of unstructured documents containing related data items on one or more topics, e.g. a text document or a TV broadcast, or more formally structured databases of facts. Intelligent knowledge management can help identify and integrate information in useful ways. A number of different content analysis technologies have been developed to either locate information within document collections either automatically or in collaboration with a human operator.

Many technologies form components of an intelligent knowledge management system. These include *information retrieval*, *summarization*, content reduction by specialization, content and context abstraction, translation of content between one or more languages; *content indexing*, speech or image recognition and analysis to identify the content of multimedia data; and *information visualization*. Content and context abstraction presented in the preceding chapters encompass *metadata* or *manual annotation* of the content, useful in identifying document contents or adding additional information beyond the content itself, identification of *context*, e.g. identifying temporal or location information associated with information, and *user interaction* with media, the platform to which it is being delivered, and user tasks. Since many of the technologies being applied have their origins in text processing technologies for natural language content, the context abstraction techniques the author has presented can be deployed. Provided that reasonably robust indexing is available, these technologies have generally been shown to transfer well to multimedia data.

A significant problem for multimedia content is the identification of the content in to start with, before the natural language processing based technologies can be applied. Multimedia content is of course potentially richer than it is possible to exploit by making use only of its linguistic components. Interpreting and exploiting linguistic content is of course challenging in its own right, but for multimedia spoken and image based linguistic data language processing made robust to indexing errors. Ideally the presence of errors should be detected and either corrected or handled in some ordered way.

A much more open question is how to index nonlinguistic visual and audio content and how to exploit this in knowledge management. More than merely retrieving and extracting data from published or carefully edited items, automatic processing of unstructured documents which may not have been formally published or archived raises the possibility of capturing and exploiting *tacit* knowledge which may not appear elsewhere (Brown et al. 2001). Tacit knowledge in this context is knowledge that appears in passing or ephemeral form, for example in discussion or extemporization during presentations. Multimedia knowledge management tools offer the possibility to record, index, and retrieve tacit knowledge.

6.3. Indexing Multimedia Content

Before the information held within a document or information stream can be utilized, it must be made available by a process of indexing. In the case of electronic text this is a relatively simple process often requiring some removal or standardization of markup, and possibly some standardization or correction of spelling. The data can then be passed to technology specific processes such as speech annotation, and parsing. For other media indexing is a much more demanding task requiring automatic recognition of the content prior to the use of information management tools. The preceding chapters are the basis to tackle the problem with new techniques.

The most widely adopted approach to indexing spoken data is to form automatic transcription using a *large vocabulary recognition* (*LVR*) system. Indexing video and image data is a much more complicated task. The hard question for which there is not a simple answer in the technologies outside the present book is, what should be indexed? Ideally all aspects of the image should be captured, but what are the essential aspects? Examples are to identify the type of location and perhaps uniquely identify it. We could attempt to recognize all the objects in the scene and go further to track their motion in the scene. How should an object to classified? Should a car be classified as a car or a vehicle, and need it include attributes such as color, manufacturer, model, etc? The KR techniques in the preceding chapters have addressed the above questions.

6.4. Information Retrieval

Once the data has been indexed, the first stage of information management is usually to use the index information to identify documents that may be of use. This is most often achieved using information retrieval (IR) techniques, which are currently most often based on best-match ranked document output. IR models were originally developed for electronic text data, but have been successfully applied to retrieval of other natural language based material, including spoken documents and scanned text document images. Content recognition errors for multimedia documents of this type inevitably impact on retrieval accuracy. The retrieval of images and video based on visual content is again more technically challenging and technically problematic. For example an initial question for an image retrieval system is how should the search query be created? Various methods have been considered to address this problem. These have included using text-based search requests to locate an image based on labels assigned to the image either manually or automatically by OCR of text in the image or labels of individuals, objects, or potentially events automatically identified in the image.

The least problems are the very high cost of assigning labels manually. That is subjective to the intended use of the image anticipated by the labeler and their knowledge of the domain. An alternative approach is to actually use images to look for images, for example, the "query-by-example" approach uses an existing image to search for similar ones, or sketching can enable the searcher to draw some likeness of the images or objects they are looking for (Carson et al. 1999). There is, however, a further issue with the retrieval of image date; it is difficult to specify what the information need actually is. It is difficult for a searcher to specify what they are actually looking for.

7. CONCLUSIONS

The overview to a multimedia language, a logic—the Morph Gentzen logic and a brief view to its models are presented. The techniques are a basis to design intelligent business interfaces, content management, ERP interfaces, and decision support systems. The knowledge bases consists of behavior descriptions, vocabulary definitions, visual objects and relations, decision rules and uncertain facts. Agent tress are applied to reach decisions. Generic diagrams are applied to an encoding for a minimal efficient KR technique applied to define relevant world models and KR for data search and discovery. Model checking is simplified to the same degree. New areas where the paper applies to is heterogeneous computing Nourani (1998) and mediation Wiederhold [7]. Proof abstraction is yet another area where KR can be keyed to KB and applied to knowledge discovery. We have defined various notions of generalized diagrams for KR applications and KB data discovery. KB completions can be applied with our techniques in Nourani [4, 6]. Multimedia datawarehousing and discovery computation techniques with predictive diagrams is briefed. There are management science areas applied to on projects with colleagues at The University of Auckland being reported at conferences (Nourani and Loo 2000). User modeling is a prerequisite to usability. [23] is a start for anyone who is trying to establish a usability presence. Topics include: making a business case for usability, overcoming inertia in large organizations, integrating usability into system development, and cultivating an effective client relationship.

Content retrieval and knowledge management has been examined as is being carried out by the current efforts and new perspectives offered with the techniques offered on the chapter. Content retrieval, business intelligence, and knowledge management areas are outlined and the difficult challenges are stated. The techniques presented in the chapter and the book are proposed as the basis to address the problem area and reconcile the challenges with specific techniques.

EXERCISES

1. What techniques and where on the business interface and business logic can agent learning techniques be applied to business intelligence to increase the competitive edge?
2. a. How do agent game trees differ from ordinary game trees?
   b. How are agent game trees applied to decisions?
   c. Can you decide if there is enough node information to proceed with a decision from agent state tables?
3. What context abstraction techniques can be applied to lift annotated hypertext to icon image indexing?
4. The preliminary design to a content processing with KM might retrieve images and video based on visual content.
   4a. What intelligent agent multimedia techniques can be applied to accomplish the task?
   4b. An initial question for an image retrieval system is how should the search query be created? Various methods have been considered to address this problem. How can you apply agent JAVA, objects, or Morph Gentzen based events to automatically identified in the image.
   4c. Design and implement part b.
5. The very high cost of assigning labels manually can be avoided with multiagent intelligent multimedia techniques. Can you specify how to actually use images to look for images.
6. While the SQL Server "Yukon" blurs the lines between relational and multidimensional databases in the above intelligent interface model fills the RDB box, and the design yet has to deploy the intelligent content IVBM TAIM.
   6a. Can you apply SQL KR techniques to query content with IVBM that can be retrieved with Morph Gentzen image comparison?
   6b. What content and context abstraction techniques are applied to realize the design?
   6c. What are example agents that you have to apply?

REFERENCES

1. J. Bailey, M. Georgeff, D. B. Kemp, D. Kinny, K. Ramamohuarao, "Active Databases and Agent Systems, A Comparison." Technical Report, University of Melbourne, Parkville, 1995.

2. G. Brajnik, G. Guida, and C. Tasso, *IEEE Transaction on Systems, Man, and Cybernetics* 20, 166 (1990).

3. M. G. Brown, J. T. Foote, G. J. F. Jones, K. S. Jones, and S. J. Young, "Open-Vocabulary Speech Indexing for Voice and Video Mail Retrieval" (P. J. Brown, Ed.), pp. 307–316. In Proceedings of ACM Multimedia, ACM, Boston, 1996.

4. D. Browne, "Studio: Structured User-interface Design for Interaction Optimization." Prentice Hall, New York, NY, 1994.
5. C. Carson, M. Thomas, S. Belongie, J. M. Hellerstein, and J. Malik, "Blobworld: A System for Region-Based Image Indexing and Retrieval," pp. 509–516. In Third International Conference on Visual Information Systems, Amsterdam, 1999.
6. W. F. Cody, L. M. Haas, W. Niblack, M. Arya, M. J. Carey, R. Fagin, D. Lee, D. Petkovic, P. M. Schwarz, J. Thomas, M. Roth, J. H. Williams, and E. L. Wimmers, "Querying Multimedia Data from Multiple Repositories by Content: The Garlic Project," Proceedings IFIP 2.6 Third Working Conference on Visual Database Systems, Chapman & Hall, Lausanne, Switzerland, 1995.
7. K. Dekker and V. Lesser, "Analyzing the Need for Metalevel Communcations." In Proc. of the Twelfth International Workshop on Distributed AI, DAI 93, Hidden Valley, Pennsylvania, 1993.
8. M. Genesereth and N. J. Nilsson, "Logical Foundations of Artificial Intelligence." Morgan-Kaufmann, 1987.
9. G. Gentzen, *Math. Ann.* 119, 140 (1943).
10. G. J. F. Jones, *Personal and Ubiquitous Computing* 5, 253 (2001).
11. G. J. F. Jones, Ed., "AAAI Spring Symposium on Intelligent Multimedia Knowledge Management," AAAI Technical Report SS-03-08, 2003.
12. G. J. F. Jones and A. M. Lam-Adesina, "Examining the Effectiveness of IR Techniques for Document Image Retrieval." In Proceedings of the Workshop on Information Retrieval and OCR: From Converting Content to Grasping Meaning, Tampere, 2002.
13. M. Mitra and B. B. Chaudhuri, *Information Retrieval* 2, 141 (2000).
14. N. J. Nilsson, "Searching, Problem Solving, and Game-Playing Trees for Minimal Cost Solutions" (A. J. Morell, Ed.), Vol. 2, pp. 1556–1562, North-Holland, Amsterdam, 1969.
15. C. F. Nourani, "Planning and Plausible Reasoning in AI," pp. 150–157. Proc. Scandinavian Conference in AI. IOS Press, Roskilde, Denmark, 1991.
16. C. F. Nourani, "Free Proof Trees and Model-Theoretic Planning." AISB, Sheffield, 1995.
17. C. F. Nourani, "AII and Heterogeneous Software Design," Eighth European Workshop on Modelling Autonomous Agents in a Multi-Agent World, University of Karlskrona/Ronneby, Ronneby, Sweden, 1996a.
18. C. F. Nourani, "Slalom Tree Computing—A Tree Computing Theory for AI" AI Communications, Vol. 9, No. 4, IOS Press, Amsterdam, 1996b.
19. C. F. Nourani, "Agent Computing, KB for Intelligent Forecasting, and Model Discovery for Knowledge Management," AAAI Wkshp on Agent Based Systems in the Business Context Orlando, Florida, 1999a.
20. C. F. Nourani, "KR and Model Discovery with Predictive Logic." AI Conference, AKA Wkshop, Sydney, 1999b.
21. C. F. Nourani, "Intelligent Multimedia—New Computing Techniques and Its Applications" (Ch. Freytag and V. Wolfengagen, Eds.). CSIT'99, Proceedings of 1st International Workshop on Computer Science and Information Technologies. MEPhI Publishing, Moscow, Russia, 1999c.
22. C. F. Nourani, "Intelligent Multimedia—Computing Techniques and Design Paradigms," Lulu Publishers, 1999d.
23. C. F. Nourani, "AII, Heterogeneous Design, and Retrieval Agents," IJCAI-99, IIIS Track, Stokholm, 1999e.
24. C. F. Nourani, "Infinitary Multiplayer Games," Summer Logic Colloquium, Utrecht, 1999f.
25. C. F. Nourani, "Intelligent Heterogeneous Multimedia and Intelligent Active Visual DB and KB." ICCIT, Las Vegas, Nevada, 2000.
26. C. F. Nourani, "The TAIM Intelligent Visual Database," 12th International Workshop on Database and Expert Systems Applications, Wkshp Records, IEEE Computer Society Press, Germany, 2001.
27. C. F. Nourani, "Multiplayer Games, ERP and Competitive Models." Fraunhofer Institute for Open Communication Systems, Germany, 2002a.
28. C. F. Nourani, "Multiagent Games, Competitive Models, and Game Tree Planning" AAAI Fall Symposium on Intent Inference. Falmouth, Massachusetts, AAAI Press, 2002b.
29. C. F. Nourani, "Virtual Tree Computing, Meta-Contextual Logic, and VR," Vol. 8. BSL, ASL Spring, Seattle WA, 2002c.
30. C. F. Nourani, "KR, Predictive Model Discovery, and Schema Completion," International Cybernetics and Systems Conference, Florida, 2003.
31. C. F. Nourani and G. S. L. Loo, "Intelligent Business Objects and Agent Computing," Parallel and Distributed Computing. University of Auckland, CSREA Press, Las Vegas, Nevada, 1998.
32. C. F. Nourani and G. S. L. Loo, "KR and Model Discovery from Active DB with Predictive Logic." Data Mining Applications to Business and Finance, University of Auckland, Cambridge, UK, 2000.
33. C. F. Nourani and R. M., "Model Discovery, Intelligent W-Interfaces, and Business Intelligence with Multitier Designs." Accepted to the International EC-conference, Cox Business School, Dallas, TX, 2003.
34. C. F. Nourani and K. Maani, "Business Modeling and Enterprise Planning: Two Synergistic Partners SSGRR," Invited paper. The University of Auckland, MSIS, L'Auquila, Rome, Italy, 2003.
35. D. Prawitz, "Natural Deduction: A Proof Theoretic Study." Almqvist and Wiksell, Stokhlom, 1965.
36. Sebastian Thrun, C. Faloutsos, T. Mitchell, and L. Wasserman, "Automated Learning and Discovery: State-of-the-Art and Research Topics in a rapidly Growing Field." CMU-CALD-98-100, Center for Automated Learning and Discovery Carnegie Mellon University, Pittsburgh, PA, 1998.
37. L. Trenner and J. Bawa, Eds., "The Politics of Usability: A Practical Guide to Designing Usable Systems in Industry." Springer-Verlag, London, UK, 1998.
38. Gio Wiederhold, "Views, Objects, and Databases," Vol. 19, pp. 37–44. IEEE Computer, 1986.

CHAPTER 12

Heterogeneous Computing, Agent-based Structures, and Implementations

CONTENTS

1. INTRODUCTION

A software design paradigm incorporating novel implementation techniques and a definition of abstract intelligent implementations (AII) are presented. Innovative techniques for design of knowledge bases and heterogeneous software systems with techniques for automated reasoning are put forth. The application areas include support for highly responsive planning. Intelligent implementation of software, i.e., the design and implementation by AII techniques is due to be an area of crucial importance. The AII techniques are being applied (Nourani 1995c) gradually to the real problems encountered in fields such as intelligent systems, aerospace, robot design, and knowledge bases, abbreviated by KB. A basis for what we have called artificial algebras (Nourani 1995b) has been written since the first version of AII was written in 1993. It is a preliminary theory for an algebra for intelligent trees and artificial intelligence. The mathematical foundations for software agents might call for algebras with varying carriers and functionality. AI systems might be defined by the stages of conceptualization, design, and implementation. Each of the stages is to be approached in ways that minimize human error and enable the designed system to automatically recover from faults. The fault recovery issues are not the topic of this paper and are treated by this author

ISBN: 1-58883-037-3/$35.00

in Nourani (1993). We design software with agents via a methodology which commences with a knowledge acquisition phase, followed by a specification phase, and concluded by a system implementation phase. The present approach defines functional nondeterministic knowledge learning (Design_Agents), fault free system specification, and multiagent abstract implementations. Design_Agents is flagrant agent computing by active agent learning, and includes exception knowledge as an essential component, as does system specification. The techniques are defined for designing heterogeneous software. System implementation is by independent concurrent computing agents. A pair of systems, each consisting, defines AI and software systems in the present paper of many computing agents. The two parts are mutually synchronized to enable fault and exception handling and recovery in an automatic manner (Nourani 1993). *Software agents* are specific agents designed by a language that carry out specified tasks and define a software functionality. Most agents defined by our examples are software agents. In the space examples there, of course, implied hardware functionality specified. Objects are in the well-known sense of the word in object programming, abbreviated by OOP. However, our designs are with *intelligent objects* a concept we had invented since 1992. Its foundations has been developed and applied in the authors publications referenced here. Ordinary objects consist of abstract data, perhaps encapsulation, and operations. Most recent programming techniques apply OOP in some form. Software engineering techniques with abstract data types have had OOP on their mind. IOOP (Nourani 1995d) is a recent technique developed by the author combining AI and software agents with OOP. For our project the modular programming concepts are combined with software agent computing, new IOOP constructs object-coobject pairs and kernels. Modules are aggregate objects with a specific functionality defined. Aggregate objects and their specified functions are defined by ⟨module-comodule⟩ pairs called *kernels*. A kernel consists of the minimal set of processes and objects that can be used as a basis for defining a computing activity. The term kernel is analogous to the terminology familiar in operating systems concepts, but is at a high level of abstraction with objects and functions encapsulated. A system is defined by a set of kernels, which can be programmed to function synchronous applying software agents. The analogy is a distributed computing environment with many computers on a net. Each kernel corresponds to a computer on the net. The multiagent AI concepts are the standard terms (Genesereth and Nilsson 1987). For the intelligent systems, the nomenclature, e.g., facilitator, mediator (Wiederhold 1993, 1994) is from standardization defined and agreed on at a conference in Colorado for the purpose in 1994–1995 [see Wiederhold (1991)]. Heterogeneous design with software agents dates back to the Nourani [1991–1992] papers and addressed on a formal funding proposals the same years. The applied terminology is defined in our paper. The same conventions define *heterogeneity* to be the mismatch found in autonomously developed resources and services, ranges from platforms, operating systems, database systems and models, data representations, ontologys, semantics, and processing paradigms.

Level is a conceptual categorization, where objects at a lower level depend on their ancestors at a higher level. Ancestor is an object at a higher level, source of inheritable attributes. The root object is the ultimate ancestor. Ontology is a set of terms and relationships used in a domain, denoting concepts and objects, often ambiguous among domains. The techniques in Nourani (1983, 1993a, 1995) have started to be applied to the design of knowledge-based systems by the present project and (Koopman et al. 1991). The basis for a sound theoretical and practical methodology for designing AI software systems is emerging. The paper's structure is as follows. Section 1 defines the way multiagent systems might be specified by software agents. Additional new concepts applied are ⟨object-coobjects⟩ and intelligent objects. There are illustrating examples. Section 2 combines the designs with abstract mediators and applies the current intelligent systems terms to define formal agent-based designs. Section 3 defines event-prompted agents. Section 4 defines multiagent systems designed with the techniques, instantiating facilitators and mediators by an example. Section 5 defines formal algebras for multiagent systems and defines formal implementation maps for the algebras. It further defines ontology algebras incorporating the ontology preservation principle. Section 6 is an overview to the AIS synthesizer for multiagent software design. The paper is concluded by Section 7.

2. SPECIFYING MULTIAGENT SYSTEMS

The hypotheses for the realization of systems in our project might appear "linear" steps of software engineering, however its linearity is no more stringent than the concept of modular design. It is the least we can demand from a design. In reality the design concept is highly nonlinear. The agents can be applied in ways which, compared to an ordinary software engineering design, appear highly nonfunctional and non-modular. From the software agent designer's viewpoint, however, there is modularity with artificial structures. Artificial structures (Nourani 1995b) are implemented by agent morphisms. The process is thus includes loops amongst the phases in the software life cycle. The intelligent objects and modules, agents, facilitator and mediators leave many degrees of freedom to design. There are artificial loops in the design resembling aerobatics by high-speed airborne agents. The AI and software designer specifies the actions and operations from the informal specifications supplied by a "user." The initial phase of the design of the proposed AII techniques is to present the design with mediator (Wiederhold 1993, 1994), Section 2. Abstract specifications, where specifications are in the sense of Nourani (1983, 1987b, 1993a). Ontology algebras are defined at meta-data and meta-knowledge level. Intelligent tree computing theories (Nourani 1994) and artificial algebras (Nourani 1995b) can be applied to the theoretical development. Knowledge acquisition requires either interviewing an expert, brainstorming with a group of experts, or structuring one's thoughts if the specifier is the expert. For multiagent designs there are active learning agents and automatic learning. We present the notion of functional nondeterministic knowledge learning (Design_Agents) in Nourani (1993). Design_Agents is formulated to deal with the conceptualization stage and is being applied by the present project to define active learning by agents. Design_Agents requires the user to inform the specifier as to the domains that are to be expected, i.e. what objects there are and what the intended actions (operations) on the objects are, while fully defining such actions and operations. The actions could be in form of processes in a system. The relations amongst the objects and the operations (actions) can be expressed by algebras and clauses, which the specifier has to present. The usual view of a multi-agent systems might convey to an innocent AI designer that an agent has a local view of the environment, interacts with others and has generally partial beliefs (perhaps erroneous) about other agents. On the surface the Design_Agents specification techniques might appear as being rigid as to what the agents expect form other agents. The Design_Agents specification does not ask the agents be specified up to their learning and interaction potential. Design_Agents only defines what objects might be involved and what might start off an agent. It might further define what agents are functioning together. Thus specifications are triples ⟨O,A,R⟩ consisting of objects, actions and relations. Actions are operations or processes. The views of abstraction (Nourani 1983, 1993a), object-level programming, and agent views of AI computation are the important components of inter-play in the present paper. Design_Agents have some additional requirements to be put forth. The requirement is that each object to be defined has to have a dual definition in terms of the actions to be taken for flagrant agents, faults, exception and recovery. At the knowledge learning phase the expert is to state all exceptions to actions and what recovery and corrective actions are to be carried out. For each action on an object a dual action is to be supplied through Desig_Agents, such that a specifier can fully define the effect of the dual actions. As an illustration the following trivial example is presented. The design techniques do not imply asking the expert to state all the exceptions to actions. The exceptions naturally present themselves by the object-coobject concept. A coobject is an object defined with the same carriers as the object, but with a set of operations complementary to the object's operations carrying on an alternate symmetric Exception operations. In the figure let OPS denote operations, EXP denote exceptions. The last equations define the exception action. In the example there is a process (action) that is always checking the supply of Angelika coffee implementing the exception function. APs:= ⟨A trivial example⟩, many robots appear at a critical entrance at once, necessitating FA activity. APs are computing events which activates an agent (see Section 3).

Object := Coffee_Constellation
OPS := Serve_Coffee (Type,Table_no) | ......

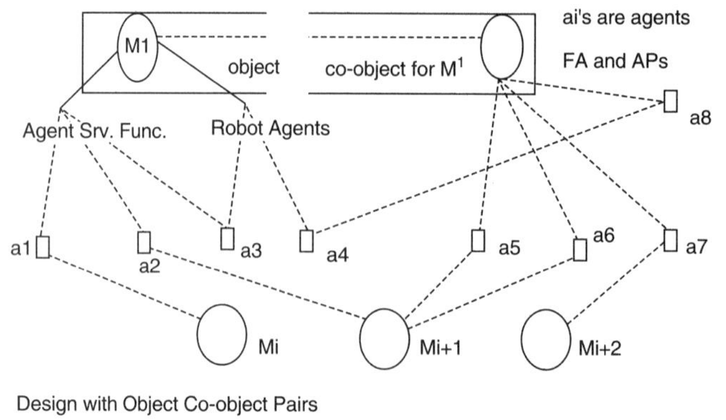

Figure 1. A heterogenous multiagent module.

Serve_Coffee (Spectacular_Brew,n)
  ⇒ Signal an available robot to fetch and serve (Spectacular_Brew,table n)
Exp := Serve_Coffee (Angelika,Table_no) | . . .
Serve_coffee(Angelika,Table_no) ⇒ if out_of_Angelika notify Table_no;
    offer cookie⟨and make use of
    intelligent decision procedures to
    offer alternatives⟩

The exception knowledge has to be acquired in every case at the Design_Agents phase. The methods noted in Nourani (1993) allow us to accomplish this in a systematic way. The following is an example mediator instantiation for the stellar robot populated coffeeshop.

In the above figure the Mi's are objects; ai's are agents. The dotted lines are agent message paths. A specific super module appears as an ⟨object-coobject⟩ pair M.

3. THE FORMAL BASIS

The present approaches have a theoretical basis abbreviated in the following sections. We start with agents, define modules and algebras, and agent and module morphisms.

3.1. Agents

Starting with what are called hysterectic agents (Genesereth and Nilsson 1987). A hysterectic agent has an internal state set I, which the agent can distinguish its membership. The agent can transit from each internal state to another in a single step. Actions by agents are based on I and board observations. There is an external state set S, modulated to a set T of distinguishable subsets from the observation viewpoint. An agent cannot distinguish states in the same partition defined by a congruence relation. A sensory function $s\colon S \to T$ maps each state to the partition it belongs. Let A be a set of actions which can be performed by agents. A function action can be defined to characterize an agent activity action: $T \to A$. There is also a memory update function mem: $I \times T \to I$. To define agent at arbitrary level of activity knowledge level agents are defined. All excess level detail is eliminated. In this abstraction an agent's internal state consists entirely of a database of sentences and the agent's actions are viewed as inferences based on its database. The action function for a knowledge level agent maps a database and a state partition t into the action to be performed by an agent in a state with database and observed state partition t.

$$\text{action: } D \times T \to A$$

The update function database maps a state and a state partition t into a new internal database.

$$\text{database: } D \times T \to D$$

A knowledge-level agent is an environment is an 8-tuple shown below. The set D in the tuple is an arbitrary set of predicate calculus databases, S is a set of external states, T is the set of partitions of S, A is a set of actions, see is a function from S into T, do is a function from A S into S, database is a function from D $\times$ T into D, and action is a function from D $\times$ T into A. $\langle$D,S,T,A,see,do,database,action$\rangle$. Knowledge level agents are hysterectic agents.

3.2. Agent Morphisms and Module Preservation

Starting with what we called hysterectic agents (Genesereth and Nilsson 1987). A hysterectic agent has an internal state set I, which the agent can distinguish its membership. The agent can transit from each internal state to another in a single step. Actions by hysterectic agents are based on I and observations. The observations are from problem solving boards, messages to the agent, and a database. There is an external state set S, modulated to a set T of distinguishable subsets from the observation view point. An agent cannot distinguish states in the same partition defined by a problem congruence relation. A sensory function $s\colon S \to T$ maps each state to the partition it belongs. Let A be a set of actions which can be performed by agents. A function action can be defined to characterize an agent activity action: T $\to$ A. There is also a memory update function. A *hysterectic* agent HA defined by a sextuple $\langle$I,S,T,A,s,d,internal,action$\rangle$ where d is a function form $A \times S \to S$ and internal $I \times T \to I$. Let HA be a set of sextuples defining a hysterectic agents. Define HA morphism by a family of functions defined component-wise on the sextuple above.

DEFINITION 3.1. A HA morphism is a function F: HA $\to$ HA$'$ defined component-wise by F[i]: I$\to$ I$'$; F[S]: S $\to$ S$'$, F[T]: T $\to$T$'$, F[A]: A $\to$A$'$; F[s]: S$\to$ T$'$; F[d]: A$'$ $\times$ S$'$ $\to$ S$'$ and F[internal]: I$'$ $\times$ T$'$ $\to$ I$'$ F[action]: I$'$ $\times$ T$'$ $\to$ A$'$. □

Definition 2.1 implies F defines a new hysterectic agents from HA by a morphism. The definition might become further transparent in view of definitions is Section 2.4. Component-wise definitions for a morphism might be viewed as functions on a multi-sorted signature carrying the sextuple. Similar morphisms can be defined for knowledge level agents defined in Section 2.1 which we can refer to by KL-morphisms.

3.3. Agents, Modules, and Algebras

The computing enterprise requires more general techniques of model construction and extension, since it has to accommodate dynamically changing world descriptions and theories. The models to be defined are for complex computing phenomena, for which we define generalized diagrams. The techniques in Nourani (1983, 1987, 1991, 1994) for model building as applied to the problem of AI reasoning allows us to build and extend models through diagrams. It required us to define the notion of generalized diagram. We had invented G-diagrams (Nourani 1987, 1991, 1993b, 1994) to build models with prespecified generalized Skolem functions. The specific minimal set of function symbols is the set with which a model for a knowledge base can be defined. The G-diagram techniques allowed us to formulate AI worlds, KB's in a minimal computable manner to be applied to agent computation. The techniques in Nourani (1991, 1994) for model building as applied to the problem of AI reasoning allows us to build and extend models through diagrams. A technical example of algebraic models defined from syntax had appeared in defining initial algebras for equational theories of data types (Goguen 1973) and our research in Nourani (1969). In such direction for computing models of equational theories of computing problems are presented by a pair (,E), where is a signature (of many sorts, for a sort set S) (Goguen 1973, Nourani 1995b) and E a set of equations. Signatures are in the same sense as key signatures in music.

DEFINITION 3.2. An s-sorted signature or operator domain is a family $\langle w, s\rangle$ of sets, f or s in S and w in S* (where S* is the set of all finite strings from S, including the empty string) call f $\langle w, s\rangle$ and operation symbol of rank w, s; of arity w, and of sort s. □

We apply multi-sorted algebras via Definition 3.2 to multiagent systems.

DEFINITION 3.3. Let be an S-sorted signatures. A-algebra A consists of a set As for each s in S (called the carrier if A of sort s) and a function

$\langle \mathrm{A}\rangle$: $\mathrm{A}s_1 \times \mathrm{A}s_2 \times \cdots \times \mathrm{A}s_n \rightarrow \mathrm{A}s$ for each $\langle w, s\rangle$, with $w = s_1 s_2 \cdots s_n$ (called the operation named by). For $\langle , s\rangle$, A As, i.e., the (set of names) of constants of sort s. □

DEFINITION 3.4. If A and B are algebras, a-homomorphism h: A → B is a family of functions $\langle hs\text{: } \mathrm{A}s \rightarrow \mathrm{B}s\rangle$ s in S that preserve the operations, i.e. that satisfy

(h_0) For $\langle , s\rangle$, the $hs(\mathrm{A}) = \mathrm{B}$;
(h_1) If, for $\langle w, s\rangle$, with $w = s_1 s_2 \cdots s_n$ and $\langle a_1, \ldots, a_n\rangle$ in $\mathrm{A}s_1 \times \mathrm{A}s_2 \times \cdots \times \mathrm{A}s_n$, then $hs(\mathrm{A}(a_1, \ldots, a_n)) = \mathrm{B}(hs(a_1), \ldots, hs(a_n))$. □

For an intelligent signature I, let T⟨I⟩ be the free tree word algebra of signature I. The quotient of T⟨I⟩, the word algebra of signature, with respect to the I-congruence relation generated by a set of equations E, will be denoted by T⟨I,E⟩, or T⟨P⟩ for presentation P.

3.4. Agents, Languages, and Models

By an intelligent language we intend a language with syntactic constructs that allow function symbols and corresponding objects, such that the function symbols are implemented by computing agents in the sense defined by this author in Nourani (1993c, 1996a). Sentential logic is the standard formal language applied when defining basic models. The language is a set of sentence symbol closed by finite application of negation and conjunction to sentence symbols. Once quantifier logical symbols are added to the language, the language of first order logic can be defined. A model for is a structure with a set A. There are structures defined for such that for each constant symbol in the language there corresponds a constant in A. For each function symbol in the language there is a function defined on A; and for each relation symbol in the language there is a relation defined on A. For the algebraic theories we are defining for intelligent tree computing in the forthcoming sections the language is defined from signatures as in the logical language is the language of many-sorted equational logic.

The signature defines the language by specifying the function symbols' arities. The model is a structure defined on a many-sorted algebra consisting of S-indexed sets for S a set of sorts. By an intelligent language we intend a language with syntactic constructs that allow function symbols and corresponding objects, such that the function symbols are implemented by computing agents. A set of function symbols in the language, referred to by AF, is the set modeled in the computing world by AI Agents with across and/or over board capability. Thus the language defined by the signature has designated function symbols called AF. The AF function symbols define signatures which have specific message paths defined for carrying context around an otherwise context free abstract syntax. A set of function symbols in the language, referred to by AF, are agents with nontrivial capability. The boards, message passing actions, and implementing agents are defined by syntactic constructs, with agents appearing as functions. The computation is expressed by an abstract language that is capable of specifying modules, agents, and their communications. We have put together the AI concepts with syntactic constructs that could run on the tree computing theories we are presenting in brief. We have to define how the syntactic trees involving functions from the AF are to be represented by algebraic tree rewriting on trees. This is the subject of the next section, where free intelligent trees are defined. An important technical point is that the for agents there are function names on trees.

DEFINITION 3.5. We say that a signature is intelligent iff it has intelligent function symbols. We say that a language has intelligent syntax if the syntax is defined on an intelligent signature. □

DEFINITION 3.6. A language $\mathscr{L}$ is said to be an intelligent language iff $\mathscr{L}$ is defined from an intelligent syntax. □

The example of intelligent languages we could present are composed from ⟨O,A,R⟩ triples as control structures. The A's have operations that also consist of agent message passing.

The functions in AF are the agent functions capable of message passing. The O refers to the set of objects and R the relations defining the effect of A's on objects. Amongst the functions in AF only some interact by message passing. The functions could affect objects in ways that affect the information content of a tree. There you are: the tree congruence definition thus is more complex for intelligent languages than those of ordinary syntax trees. Let us define tree information content for the present formulation. Hence there is a new frontier for a theoretical development of the ⟨O,A,R⟩ algebras and that of the AII theory. ⟨O,A,R⟩ is a pair of algebras, ⟨Alg[A],Alg[F]⟩ (see Section 3), connected by message passing and AII defines techniques for implementing such systems. To define AII we define homorphisms on intelligent signature algebras.

DEFINITION 3.7. An I-homorphism is a homorphism defined on algebras with intelligent signature I. □

To define agent specific designs we apply HA-morphisms via the following definition.

DEFINITION 3.8. Let A and B be I-algebras with signatures containing an agent signature HA. A HA-homorphism from A to B is an I-homorphism with defined HA-morphism properties. □

4. MULTIAGENT IF DESIGNS AND MEDIATORS

The term "agent" has been recently applied to refer to AI constructs that enable computation on behalf of an AI activity. It also refers to computations that take place in an autonomous and continuous fashion, while considered a high-level activity, in the sense that its definition is software and hardware, implementation, independent. For example, in a planning problem for space exploration, an agent might be assigned by a designed flight system (Nourani 1992) to compute the next docking time and location, with a known orbiting spacecraft. Agents are in most cases informable (Genesereth-Nilsson 1987), thus allowing message passing actions. We can define AII software systems designed by AI methods as intelligent agent architectures, with external behavior that is a function of the degree of message passing actions and parallelism conceptualized. Since our specifications consist of objects, actions, and relations defining the effect of actions on objects, we can define formal IF systems from the specifications and prove the specifications can be implemented by a set of agents. A *mediator* is a software module that exploits encoded knowledge about certain sets or subsets of data to create information for a higher layer of applications and the definition goes on to state 'It should be small and simple, so that it can be maintained by one expert or, at most, a small and coherent group of experts.' Mediator instantiation is to populate a domain-independent service or tool with domain-specific knowledge. We define *Mediator Specifications* consisting of a tuple of functions and relations of the form ⟨O,(A,F),(RNA,RFA)⟩, where A is actions and F computes flagrant agents from APs to faults. (RNA,RFA) are their respective relations, NA for normal action and FA for flagrant or fault actions. In the example of the last section O is Coffee_shop, and serve-coffee an example of an action, a member of A. EXP defines the set F. The third line defines an example of a relation in RNA, and the last function is an example of a relation in RFA. We had invented a twin-engine agent-based computing system. ⟨A,F⟩ := ⟨Design_Agents,CoAgents⟩, consisting of Design_Agents := ⟨O,A,RNA⟩ and CoAgents := ⟨O,F,RFA⟩. The design is depicted by the following figure.

Object-coobject pairs are applied design the $\langle A_i, F_i\rangle$ and composed to ⟨A, F⟩. The modules are defined from multiple objects. The Design_Agents corresponds to an algebra Alg[A] of normal activities and CoAgents to an algebra Alg[F] for flagrant agent computing, faults, recovery, and revision of actions. It consist of a pair of complex algebras, connected only by agent message passing.

DEFINITION 4.9. A system is intelligent fail-safe, abbreviated by IF, when defined by a pair ⟨Alg[A], Alg[F]⟩ where A and F are I-algebras where I is an intelligent signature bearing agent functions. □

Having defined the intelligent algebras, HA morphisms, and IF designs, we can define formal multiagent implementations for IF systems applying HA-homomorphism and formal

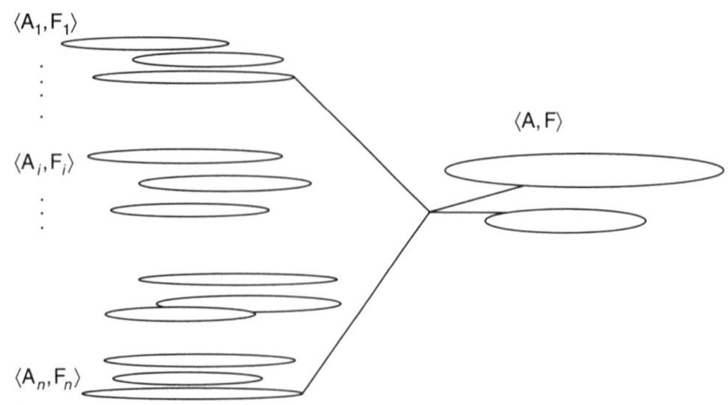

Figure 2. Multiagent object enterprise design.

implementation techniques (Nourani 1980, 1993), (Ehrig et al. 1980–1987). It is obvious how to define AII implementations direct from HA-homomorphims applied to our 1980's papers. The details are outside the scope of the present paper. Each of the Design_Agents and CoAgents consists of agents that are mutually, often pair-wise, informable. The systems ⟨Ai,Fi⟩, each consist of objects, actions and relations. Actions could be in form of operations or message communication from one object to another. A set of computing agents forms Design_Agents and a dual set forms CoAgents. Thus a pair of systems is defined that can be implemented by agents that logically or physically can be thought of as running on several microprocessors. The algebras Alg[A] and Alg[F] define wrappers for the mediators as functions for interacting with resources. A wrapper is a tool to access known resources and translate their objects.

The spontaneity and fault tolerance degree is a function of the intelligence of the agents implementing the ⟨Design_Agents,CoAgents⟩ pair. The agents have incomplete information about the immediate needs of activating other agents or exceptions. Thus the efficiency and strength of functionality of our software systems are a function of the degree of intelligence we build in the implementing agents. The agents must have some reasoning ability to at least make transition or message passing decisions.

This approach allows us to design systems that can deal with unplanned or erroneous behavior in an AI system. The next step is defining the ⟨Design_Agents,CoAgents⟩ from the flagrant agent knowledge learning (Design_Agents) inputs. Its implementation consists of an autonomous pair of communicating systems to be defined in the following section. We have thus defined a formal computing model, the IF definition consisting of an algebra of processes and objects, with possible use of new parallel languages and intelligent object programming put forth in preliminary reports by this author (Nourani 1992, 1997). Theories for intelligent syntax tree computing are being put forth by this author in Nourani (1996). Starting form our techniques, programs capable of generating mediators, routers, and translators from formal specifications can be designed. In some cases these generators may work automatically, in some cases interactively with humans.

An *agent provocateur*, henceforth abbreviated by AP, is a computing event which activates an agent, specifically a fault agent. There are many sorts of APs and they affect ontologys in ways unknown to the AP and to its originating resource, its content, model, and domain. Thus APs necessitate an ontology update. Accommodating the update is a challenging problem. The software agent system designer has to define APs. APs might also be defined by learning agents. Their structure is design specific and run by the agents active with the events. A computing system with predefined APs was experimented with in Nourani (1993) and an AII model was presented in the author's projects, where APs were all defined in advance. The terminology AP is new to the present paper however. When the APs are not defined and known in advance the ontology revision problem is quite difficult. Defining mediators by abstract multiagent specifications and AII are ways to manage APs. Section 5 defines specific APs.

5. AI AND CONCURRENT AGENT PROCESSES COMPUTING

The above multi-agent implementation of the mediator specifications implies design with a pair of concurrent systems. Each of the two systems is to be designed with a collection of modules, such that there corresponds a module for each specification. A module consist of the minimal set of processes and objects that can be used as a basis for defining a computing activity. The objects and the operations of one set of modules once defined specifies the basis for Design_Agents, while those of the CoAgents' basis is defined by the dual module. The set of modules defining Design_Agents and CoAgents are synchronized by cross operations and interact by some operations that are implemented by message communications between Design_Agents and CoAgents. These operations are defined to either inform the various processes that are mutually dependent or to take the system from an active state in Design_Agents to an active state in CoAgents. Note that when exceptional conditions occur the active state is CoAgents. However, both sets of modules are considered concurrently "running."

CoAgents' major task is that of handling unexpected events, recovery from faults, and revision of actions. Thus CoAgents has to know what agents can become active to compute for APs and be designed to activate remedies for ontology revision. If exception recovery takes place, in each module, the active module (a collection of agents) for a particular function, will be the Design_Agents' component, while the CoAgents component does concurrent checks for further exceptions should they be encountered. In each of the modules there are objects, processes defining the operations, and objects to which there is a corresponding function in the other module. Thus Design_Agents and CoAgents imply a set of objects and processes defined by many-sorted OAR algebras. The objects O_1 are many-sorted structures with the $\langle p_i \rangle$, $\langle q_i \rangle$ and $\langle e_i \rangle$ as the operations. RNA and RFA define the algebras via relations.

$$\text{Design\_Agents} := \langle \{O_1, \langle p_1, \ldots, p_n \rangle\}, \{O_2, q_1, q_2, \ldots\}, \ldots \{O_n, \ldots\}, \text{RNA} \rangle$$

RNA is the set of relations on each object and cross objects.

$$\text{CoAgents} := \langle \{O_1, \langle e_1, \ldots, e_n \}\rangle, \{O_2, \langle e_{11}, e_{12}, \ldots, e_{1m} \rangle\}, \ldots, \{O_n, \langle \ldots \rangle\}, \text{RFA} \rangle$$

RFA is the set of relations on each objects and cross objects.

Each of the processes can have a corresponding agent in the dual family. The ⟨Design_Agents,CoAgents⟩ pair in a computing system "run" as a concurrent family of processes. Various functions in Design_Agents and CoAgents are represented by agents that are mutually informable across the ⟨Design_Agents,CoAgents⟩ pair. For the fault model there is a predefined AP set and a corresponding functionality. The overall functionality of the system depends on the messages passed across from one agent to another. To each specification defined by Design_Agents there corresponds two modules running concurrent. The modules are defined by mediator instantiation for standard functioning and APs computing. There are facilitators shared by the mediators to accommodate heterogeneity. A facilitator is a component which routes messages and queries. The essential distinction between a mediator and a facilitator is one of automation and hence dynamics versus human responsibility. The vision underlying mediators is one where domain experts, equipped with inter-operating software modules, provide value-added services for data access and processing over networks. The vision underlying facilitators is 'one in which any system (software or hardware) can interoperate with any other system, without the intervention of human users or their programmers' interoperability is the capability to interoperate, often used at the transport layer. Figure 3—designs and implementation with multiagent Design_Agents, Objects-Coobjects, mediators, and facilitators. Modules 1-n define a kernel. Two kernels are mediated and there is facilitator routing the mediated computations. The pilots are wearing wrappers.

Both mediators and facilitators can provide control and routing, content manipulation, and format conversion. Facilitators and mediators can interact within an information system: Facilitators may call on mediators to perform value-added tasks. Example of a facilitators in our designs are routing agent messages between the two system parts, e.g., formally between

```
⟨Module 1⟩ ........................................................ ⟨Module n⟩
    |                                                                  |
                                  |
MEDIATOR  Modules                          MEDIATOR Modules |
                    +-------------------+ | +------------+
                                  |
```

```
                          FACILITATORS

            +------------+ | +------------+---------------+
MEDIATOR  Modules                          MEDIATOR Modules
        |                                          |
   ⟨Module k⟩ ..........................................⟨Module m⟩
```

Figure 3. Heterogenous design with facilitators and mediators.

the structures A and F. On the above figure it is obvious there is a facilitator for routing agent messages between two kernels. A trivial example for the coffee-shop example is routing agent messages form the normal functions and fault and exception functions.

6. ABSTRACT IMPLEMENTATION BY COMPUTING AGENTS

The term implementation here refers to the process of transforming an abstract characterization of an AI or software system to concrete representations and executable code. Thus implementations express the relationship between two forms of representations. The notion of abstract implementation defined by this author in Nourani (1983, 1993a) are either algebraic or model-theoretic (algebraic logic) definitions. We refer to specifications of the form ⟨O,A,R⟩ as presentations. We also expect a presentation of the form ⟨I[O],I[A],I[R]⟩ for the implementing abstract or concrete machine. The former could be the designer's conceptualization, and the latter the specification of the syntax and semantics of a programming language. This is similar to how the problem was viewed by our group in 1978–1980's, and there were many research papers that were developed by us and Ehrig et al. (1980) for the most part.

Informally the process of implementation was defined by this author to be that of encoding the algebraic structure of the conceptualization of a problem onto the algebra that specified an implementing machine (a programming abstract machine). Thus the problem was that of defining such implementations by morphisms of algebras. The problems we are proposing are to address are much more complex. It is because the implementations proposed for AI systems are by multiagent designs. Each of the functions defined by ⟨O,A,R⟩ are implemented by agents, that characterize the implementation function

$$\text{I: } \langle O,A,R\rangle \rightarrow \langle I[O],I[A],I[R]\rangle \quad \text{is to be defining a mapping}$$

$$\text{I: } \langle \mathrm{Alg}[A],\mathrm{Alg}[F]\rangle \rightarrow \langle \mathrm{Alg}[I(A)],\mathrm{Alg}[I(F)]\rangle$$

We refer to Alg[A] and Alg[F] are what we call *ontology algberas*. The implementation mapping I defines wrappers to resources in a manner preserving the ontology algebra. Ontology algebras are multi-sorted algebras defining multiagent systems defined by formal agents, e.g., hysterectic or knowledge level agents and agent morphisms. A formal definition is provided in Section 7.1. *The Ontology Preservation Principle* The AII is correct only if it preserves the ontology algebras. It will be abbreviated by AIIOPP.

Wiederhold's domain knowledge base algebra DKB consists of matching rules linking domain ontologys. There are three operations defined for DKB. The operations are Intersection—creating subset ontology and keeping sharable entries. Union—creates a joint ontology merging entries. Difference—creates a distinct ontology and removing shared entries. Mapping functions must be shown to preserve ontologys. Applying AIIOPP we can state specific preservation principles as follows. The DKB preservation Principle-AII implementations must preserve ontologys under intersection, union, and difference operations.

7. AGENT ONTOLOGY PRESERVATION THEOREMS

Let us apply the definition for HA agents and HA morphisms to state a preservation theorem. Let A and B be I-algebras with the signature I containing HA agents. Let Alg[B] be an I-algebra defined from B implementing (Nourani 1983, 1993a, 1995; Ehrig et al. 1980) a specified functionality defined by A. An AII is an implementation for Alg[A] by Alg[B].

DEFINITION 7.10. Let A and B be I-algebras with intelligent signature I containing agents. An I-ontology is an I-algebra with axioms for the agents and functions on the signature.

THEOREM 7.1. *Let A and B be I-algebras with the signature I containing HA agents. The AII with HA morphisms defined from A to B preserve I-ontology algebras iff defined by HA-homorphisms.*

PROOF. The definition for ontologies, HA morphism, Definition 2.7 and 2.8, I-algebras and I-homorphisms entail the I-ontology axioms are preserved iff agents are carried by HA-homorphisms from A to B. □

THEOREM 7.2. *Let A and B be I-algebras with the signature I containing KL agents. The AII with KL morphisms preserve I-ontology algebras iff defined by KL-homorphisms.*

PROOF. Similar to 7.1. □

There are precise statements for preservation principles and mappings in Nourani et al. (1995). DKB mappings are specific AII's were the ontology algebra operations are the same at source and target. We prove in Nourani et al. (1995) DKB mappings are AIIOPP consistent.

REFERENCES

1. Cameron and Ito, "Grammar-based Definition of Meta-programming Systems," Vol. 6, No. 1. ACM Toplas, 1984.
2. J. de Bakker, "Mathematical Theory of Program Correctness," International Series in Computer Science, Prentice Hall, 1982.
3. H. Ehrig, H. J. Kreowski, and P. Padawittz, "Algebraic Implementation of Abstract Data Types, Concepts, Syntax, Semantics, and Correctness," Vol. 85. Proc. 7th International Colloquium on Automata, Languages, and Programming, Lecture Notes in Computer Science, Noordwijkerhout, Springer-Verlag, New York, 1980.
4. H. Ehrig, H.-J. Kreowoski, and H. Webber, "Algebraic Specification Scheme for Data Base Systems." Han-Meitner-Institut Berlin and Technical Universitat Berlin, HMI-B 266, 1978.
5. M. R. Genesereth and N. J. Nilsson, "Logical Foundations of Artificial Intelligence," Morgan-Kaufmann, 1987.
6. J. A. Goguen, J. W. Thatcher, and E. G. Wagner "An Initial Algebra Approach To Specification, Correctness and Implementation of Abstract Data Types" (R. Yeh, Ed.), Vol. IV, pp. 80–149. In Current Trends in Programming Methodology, Prentice-Hall, Englewood-Cliffs, NJ, 1978.
7. J. A. Goguen, J. W. Thatcher, E. G. Wagner and J. B. Wright, "A Junction Between Computer Science and Category Theory (parts I and II)," IBM T. J. Watson Research Center, Yorktown Heights, NY, 1975.
8. M. R. J. Koopman, J. W. Spee, W. Jonker, L. Montero, K. O'Hara, M. Mepham, E. Motta, and L. J. in't Veld, "Vital Conceptual Modelling," Vital Deliverable DD213. PTT Research, Groningen, 1991.

9. C. F. Nourani, "A Model-Theoretic Approach to Specification, Extension, and Implementation," Vol. 83, pp. 282–297. 4th International Symposium on Programming, Lecture Notes in Computer Science, Springer-Verlag, Paris, 1980.
10. C. F. Nourani, *J. Association Computing Machinery* 30, 343 (1983).
11. C. F. Nourani, "Equational Intensity, Initial Models, and Reasoning in AI: A Conceptual Overview" (Tim O'Shea, Ed.). Proc. European Conference in AI, Pisa, North-Holland, Italy, 1984.
12. C. F. Nourani, "Software Specification and Meta-Programming." International Software Engineering Conference, Proceedings Workshop on Software Specification and Design, Monterey, CA, 1987a.
13. C. F. Nourani, "CatLog: A System for Functional Realization of Abstract Module Specifications in Prolog," Proc. 20th HICSS, Kona, HI, 1987b.
14. C. F. Nourani, "Parallel Module Specification" ACM Sigplan, 1990.
15. C. F. Nourani, "A Multiagent Approach to Software Fault Tolerance." FLAIRS-93, Florida AI Conference, 1991a.
16. C. F. Nourani, "Planning and Plausible Reasoning in Artificial Intelligence: Diagrams, Planning, and Reasoning," Proc. Scandinavian AI Conference, IOS Press, Denmark, 1991b.
17. C. F. Nourani, "Multiagent AI Techniques for Validation and Verification." ACM Sigsoft Software Engineering Notes, 1992.
18. C. F. Nourani, "Abstract Implementation Techniques for Artificial Intelligent Systems by Computing Agents—A Conceptual Overview," Proc. SERF-93, Orlanda, FL, 1993a.
19. C. F. Nourani, "Double Vision Computing." Proc. IAS-4, Karlsruhe, Germany, 1993b.
20. C. F. Nourani, "Slalom Tree Computing," AI Communications, IOS Press, Amsterdam, 1994.
21. C. F. Nourani, "Intelligent Languages—A Preliminary Syntactic Theory" (Jozef Gruska and Jiri Zlatuskanbsp, Eds.). 23rd International Symposium, Mathematical Foundations of Computer Science, Lecture Notes in Computer Science, Springer, Brno, Czech Republic, 1995a.
22. C. F. Nourani, "Artificial Algebras," ACM SIGSAM, 1995b.
23. C. F. Nourani, "AII and Heterogenous Software Design," MAAMAW'97, Eighth European Workshop on Modelling Autonomous Agents in a Multi-Agent World, University of Karlskrona/Ronneby, Sweden, 1995c.
24. C. F. Nourani, "The IOOP Project, Preliminary Brief," Vol. 30, No. 2, pp. 56–54. Sigplan Notices, 1995d.
25. C. F. Nourani, "Slalom Tree Computing—A Computing Theory For Artificial Intelligence," Vol. 9, No. 4. AI Communication, IOS Press, Amsterdam, 1996.
26. C. F. Nourani, "Modelling, Validation, and Hybrid Design of Intelligent Systems" (Dieter Fensel, Ed.). Proc. KMEL98, 8th Workshop Knowledge Engineering Methods and Languages, Karlsruhe University, Karlsruhe, Germany, 1998b.
27. C. F. Nourani and K. J. Lieberherr, "Data Types, Direct Implementations, and Knowledge Representation," Vol. II, pp. 233–243. Proc. 19th HICSS, Honolulu, Hawaii, 1986.
28. C. F. Nourani, "AII and Data Algebras." 1995.
29. J. W. Thatcher, E. G. Wagner, and J. B. Wright, "Data Type Specifications: Parameterization and the Power of Specification Techniques," pp. 119–132. Proc. Tenth ACM Sigact Symposium on Theory of Computing, San Diego, CA, 1978.
30. Gio Wiederhold, "The Roles of Artificial Intelligence in Information Systems," pp. 38–51. In ras Methodologies for Intelligent Systems, Lecture Notes in AI, Springer, 1991.
31. Gio Wiederhold, "Mediation in the Architecture of Future Information Systems" (Hurson et al., Eds.), Vol. 25, No. 3, pp. 38–49. IEEE Computer Magazine, IEEE Press, 1993.
32. Gio Wiederhold, "Interoperation, Mediation and Ontologies," Vol. W3, pp. 33–48. Proc. Int. Symp. on Fifth Generation Comp Systems, ICOT, Tokyo, Japan, 1994.
33. Gio Wiederhold, "Mediation to deal with Heterogenous Data Sources," pp. 1-16. Proc. Interoperating Geographic Information Systems, LNCS 1580, Springer-Verlag, Zurich, 1999.

CHAPTER 13

Abstract Planning, Models, and Virtual Trees

CONTENTS

1. INTRODUCTION

The present section provides an overview of planning with free proof tree computation on diagrams. An algebraic definition and theory for artificial intelligence was put forth by the present author in the 1980's (Nourani 1984, 1991). In Nourani (1991) a particular approach to belief revision computation based on ordinal conditional rankings of propositions, applying the method of generalized plausible diagram was presented. It was also shown that a

ISBN: 1-58883-037-3/$35.00

Intelligent Multimedia Computing Science
By Cyrus F. Nourani
Pages: 165–176

similar ordinal argument with plausible diagrams could be applied for world model revision. World model revision computation is one of the difficult aspects of the nonmonotonic reasoning Artificial Intelligence (1980) systems, and Nourani (1991) presented a solution to it leading us to the present formulation, where models are implicitly revised by dynamic changes to the diagrams from which the models are built. What are plans: There is a universe defined for an AI problem. There is a goal formula that is to hold true in the universe starting with some premises. A plan is a sequence of operations in the universe that could result in terms that instantiate the truth of the goal formulas in the universe. That's what goes on as far as the algebra of the model is concerned. It is a new view to planning prompted by our method of planning with GF-diagrams and free Skolemized trees. It is a model-theoretic view. Proof-theoretically a plan is the sequence of proof steps that yields the proof for the goal formula. The proof theoretic view is what the usual AI literature presents.

Planning in hierarchy or abstraction spaces has interested AI since the last decade (Sacerdti 1974, Wilkins 1984).

The paper starts with presenting *Computable World Models with diagrams*. Planning with GF-diagrams is defined. Generic diagrams allow us to define canonical models with specific functions. The techniques take us to planning models as foundation. For the free proof trees defined for a goal formula from the GF-diagram there is a standard model satisfying the goal formulas. It is the model definable by the generic diagram. Generalized proofs and Hilbert Models are presented for plan proof trees when there are existential goal formulas. Ordinal planning with GF-diagrams is briefed from our 1986–1991 papers as specific plans.

GF plans and proof abstraction applications are alluded to and a Skolemized virtual proof tree planning algorithm is presented. Visual planning is new applications with Morph Gentzen VR, intelligent trees and Spatial KR. Spatial information and Morph Gentzen model functions and visual objects are presented in brief.

2. COMPUTABLE WORLD MODELS

In order to understand the overall direction certain model-theoretic issues need to be covered. In proving Godel's completeness theorem, one proceeds by constructing a model directly from the syntax of the given theory. This structure is obtained by putting terms that are provably equal into equivalence classes, then defining a free structure on the equivalence classes. For example, $f(a, b)$ is defined to be the formal string "fab." Constant symbols are added to the language to instantiate existential formulas. The reasoning enterprise requires more general techniques of model construction and extension, since it has to accommodate dynamically changing world descriptions and theories. The techniques in Nourani (1983), Nourani (1991), Nourani and Hoppe (1994) for model building as applied to the problem of AI reasoning allows us to build and extend models applying generalized diagrams. Such diagrams are used to build models with a minimal family of generalized Skolem functions.

The problem of model building is not Skolemization in the sense of theorem proving. It is similar to presenting a set of functions, in the same spirit as Godel operations for set theory, for defining the contractible universe, to define a model for a computing theory. The GSF's were defined by this author around 1981–1982 to be the functions with which initial models are defined inductively. The minimal set of function symbols are those with which a model can be built inductively. We focus on such models, since they are Initial in a category of models in the same sense as algebras are initial in Goguen (1973) and the models are computable (Nourani 1980). The set of all possible worlds may not be all of interest, for some may not be computable or not be a model of real world AI computations. Thus we must further develop the notion of a computable AI world. This task is being carried out by our papers elsewhere and by forthcoming papers of the present project.

3. AI WORLD DIAGRAMS

Belief revision and its computation are important concepts in designing plans for robots and other applications of planning and reasoning in AI. In order to formulate and solve revision computing problems within the present formulation we had defined generalized diagrams (Nourani 1987, 1991). Such diagrams are essential for the task of model building to accommodate belief revision. Rather than giving a formal definition we present the basic idea, and refer the reader to the cited references for the formal notions. The diagram of a structure is the set of atomic and negated atomic sentences that are true in that structure. The generalized diagram (G-diagram) is a diagram in which the elements of the structure are all represented by a minimal family of function symbols and constants, such that it is sufficient to define the truth of formulae only for the terms generated by the minimal family of functions and constant symbols. Such assignment implicitly defines the diagram. It allows us to define a canonical model of a theory in terms of a minimal family of function symbols. Such formulation allows us to refer to models of AI worlds in which all constant symbols in the AI world language have assignments in an AI world model. We are able to present fully defined models in our formulations, because our definition of models by a minimal set of Skolem terms allows us to have an onto map of all symbols to terms and individuals in an AI world model.

Let us review the example of a generalized diagram for the blocks world problem solver. The special functions in the generalized diagram are similar to CONS's in terms of which all objects are represented. Revising the diagram when theory revision occurs generates a new canonical model of the world. New axioms are added, and some old axioms are removed, letting us reconstruct a new canonical model of the world directly from the diagram.

The blocks-world problem solver has one diagram function:

$$\text{put: blocks,blocks,table} \to \text{blocks} \times \text{blocks,table}$$

forming a set of ordered pairs indicating the relative block positions. There are three block names A, B, and C. For example,

$$\text{put}(A, B, \text{Tab}) \text{ is defined to be } \{(A, B), \text{Tab}\},$$

where Tab is a variable representing the table configuration The top operation is defined by

$$\text{top: blocks,blocks,table} \to \{T, F\}$$

In the above table is the set of table configurations is represented by Tab. For example, top(A,B,{(C,B),(B,A)} = T, where {(C,B),(B,A)} is a table configuration. Let us suppose we represent the table as a Cons'd list of ordered pairs such as (A,B) denoting that A was put on B. Thus {A,(B,C)} is a possible table configuration in which A and B are free, but B is on C. Such table configuration is obtained by put(B,C,(A,B,C)).

The diagram has to assign truth values to top(x, y, z), where x, and y are block names, and z is a table configuration. Let Tab be the initial table configuration {A,B,C}. For example, one can define the G-diagram for the blocks-world by

$$\begin{aligned} \text{top}(x, y, z) &= \text{T}, \text{ if } z = \text{put}(x, y, \text{Tab}), \text{ or } z = \text{put}(x, w, \text{put}(w, y, \text{Tab})); \\ &= \text{F, otherwise} \end{aligned}$$

where x, y, w are variables assigned from {A,B,C} and z is a variable representing table configurations.

4. PLANNING WITH GENERIC FREE SKOLEMIZED-DIAGRAMS

4.1. Nondeterministic Diagrams

Possible worlds consists of a certain completeness property: for any proposition p and world w, either p is true in w or not p is true in w. It was an important point of Nourani (1991), Nourani and Hoppe (1994) that possible worlds can be represented by exactly the information contained in a generalized diagram. Let W be the set of all worlds and p be a proposition. Let $[p]$ be the set of worlds in which p is true. We call $[p]$ the truth-set of P.

Propositions with the same truth-set are considered identical. Thus there is a one-one correspondence between propositions and their truth sets. Boolean operations on propositions correspond to set-theoretic operations on sets of worlds. A proposition is true in a world if and only if the particular world is a member of that proposition. The approach in Nourani (1991) allows us to encode planning problems within generalized diagrams. It also allows us to carry out revision problems by assigning ordinals to diagrams. Truth and belief revision is by ordinal arithmetic on diagrams (Nourani 1991).

The approach in Nourani (1991) connected truth maintenance and world revision with plausible diagrams to the approach in Sphon (1988), Nourani (1987, 1991), and embedded modalities within model-theoretic considerations. Thus a computational encoding for possible worlds with G-diagrams had been initiated by the present project. A possible world may be thought of as a set of circumstances that might be true in an actual world. In the possible world approach rather than considering individual propositions, one focuses on the "*state of affairs*" that are compatible with what one knows to be true, being regarded as possible, relative to a world believed to be true, rather than being absolute. For example, a world w might be a possible alternative relative to w', but not to w''.

The notion of generalized diagram is extended (G-diagram) to include plausibility and nondeterminism for planning and for representation of possible worlds. An extended notion of G-diagram can encode possible worlds to capture the "maximally complete" idea and can be used for model revision and reconstruction. By assigning a plausibility ranking to formulae one can set a truth limit ordinal t as the truth threshold. Formulae with plausibility ranking less than t would be assigned 'T' and the other formulae would be assigned 'F.' Thus Nourani (1991) defined the notion of a plausible diagram, which can be constructed to define plausible models for revised theories. In practice, one may envision planning with plausible diagrams such that certain propositions are deliberately left indeterminate to allow flexibility in planning. In Nourani (1991) nondeterministic diagrams were defined by assigning an undefined "X" symbol to predicates in the diagram whose truth-values are not known at each stage of planning.

Such extensions to the usual notion of diagram in model theory are put forth in Nourani (1991). That approach was one method of avoiding the computational complexity and computability problems of having complete diagrams. Diagrams with "X" values, however, have some implementations implications in planning—that of requiring backtracking through some plans to assign the needed value of a proposition in the diagram to its actual truth value, rather than what was assumed by the planning based on the "spare of the moment" decision mechanism. This is not always the desired method of implementing plans. It was put forth as one method of avoiding the complexity problems of incomplete diagrams.

4.2. Generalized Free Skolem Diagrams

In the present paper we devise new techniques to achieve nondeterminism by proposing what we shall refer to as *Free Skolemized Diagrams*. Such diagrams are defined by assigning truth values to predicates with known truth values at each stage, and by assigning generalized Skolem functions (Goguen 1973, Nourani and Hoppe 1994) to the predicates whose truth values are not known. Hence we have define diagrams in which predicates are replaced by their corresponding Skolem functions. For example in a blocks world planning problem, the entry in the diagram function for top(A,B) predicate is replaced by a Skolem function F_top(A,B). We refer to such diagrams as GF-diagrams.

A put (A,B,Tab) operation in an implementation of such plans, also has the effect of setting F_top(A,B) equal or congruent to T for true. Thus any proof tree for such planning problems can in effect reduce F_top(A,B) to T. Note that the GF-diagrams as in the approach in Nourani (1991), Nourani and Hoppe (1994), also include all the information required to define possible worlds. Absolute formulae can be assigned plausibility ranking which represent their truth-value through the belief function limit ordinal. Relativised truth is captured by altering the truth limit ordinal, and by leaving plausibility degrees free by assigning generalized Skolem functions for the unknown predicates. Truth maintenance and model revision can all be done by a simple reassignment to the diagram. The canonical model of the world is defined directly from the diagram.

4.3. Planning

The planning process at each stage can make use of GF-diagrams by taking the free interpretation, as tree-rewrite computations, for example, of the possible proof trees that correspond to each goal satisfiability. There are several ways in which such planning can be achieved. The method that is proposed here is to make use of the free Skolemized proof trees in representing plans in terms of generalized Skolem functions. There are some techniques that are made use of in some recent AI literature (Hoppe 1994) which advocate the notion of partial deduction. We propose methods that can be applied to planning with GF-diagrams such that similar effects to partial deductions can be achieved and perhaps some of the techniques can be applied to implement such planning techniques. As we defined in the above sections, a plan is a sequence of operations in the universe that could result in terms that instantiate the truth of the goal formulas in the universe. Thus a tree can represent a plan. In planning with GF-diagrams the plan trees involving free Skolemized trees are carried along with the proof tree for a plan goal. The idea is that if the free proof tree is constructed then the plan has a model in which the goals are satisfied. The model is the initial model of the AI world for which the free Skolemized trees were constructed. Thus we the following free proof tree sound computing theorem.

4.3.1. Initial Models

Godel's completeness theorem proved Henkin style, proceeds by constructing a model directly from the syntax of the given theory. The structure is obtained by putting terms that are provably equal into equivalence classes, then defining a free structure on the equivalence classes. The computing and reasoning enterprise require more general techniques of model construction and extension, since it has to accommodate dynamically changing world descriptions and theories. Our techniques for model building as applied to the problem of AI reasoning allows us to build and extend models through diagrams. This requires us to focus attention on the notion of generalized diagram. The minimal set of function symbols are those with which a model can be built inductively. We focus our attention on such models, since they are Initial (Goguen 1973) and computable (Nourani 1980). We apply the canonical models since they are initial and computable.

A technical example of algebraic models defined from syntax had appeared in defining initial algebras for equational theories of data types (Goguen 1973–1976) and our projects since 1978. In such direction for computing models of equational theories of computing problems are presented by a pair (Σ, E), where Σ *is a signature* (of many sorts, for a sort set S) and E a set of Σ-equations. Let T⟨Σ⟩ be the *free tree word algebra of signature* Σ. The quotient of T⟨Σ⟩, the word algebra of signature Σ, with respect to the Σ*_congruence* relation generated by E, will be denoted by T⟨Σ, E⟩, or T⟨P⟩ for presentation P. T⟨P⟩ is the "initial" model of the presentation P. The Σ-congruence relation will be denoted by ≡P. One representation of T(P) which is nice in practice consists of an algebra of the canonical representations of the congruence classes. This is a special case of generalized standard models defined here.

In what follows $gt_1 \dots t_n$ denotes the formal string obtained by applying the operation symbol g in Σ to an n-tuple t of arity corresponding to the signature of g. Furthermore, gC denotes the function corresponding to the symbol g in the algebra C.

DEFINITION 4.1. Let Σ be an S-sorted signature. Then a canonical term Σ-algebra (Σ-CTA) is a Σ-algebra C such that

(1) |C| is a subset of T⟨Σ⟩ as S-indexed families
(2) $gt_1 \dots t_n$ in C implies t is in C and; $g_C(t_1, \dots, t_n) = gt_1 \dots t_n$.

For constant symbols (2) must hold as well, with $g\mathrm{C} = g$.

THEOREM 4.1. *Let* C *be a* Σ*-algebra. Let* P = (Σ, E) *be a presentation. Then* C *is* Σ *isomorphic to T*⟨P⟩, *iff*

(i) C *satisfies* E;
(ii) $g_C(t_1, \dots, t_n) \equiv gt_1 \dots t_n$

Note: (ii) must also hold for constants with $g_C = g$; ≡refers to the Σ-congruence generated by E.

PROOF. see Goguen et al. (1978). □

Canonical models from models to set theory had been stated for arbitrary structures as follows.

DEFINITION 4.2. *Let* $(M, a)_c$ *in* C *be defined such that* M *is a structure for a language* $\mathcal{L}$ *and each constant* c *in* C *has the interpretation* a *in* M. *The mapping* $c \to a_c$ *is an assignment of* C *in* M. *We say that* $(M, a)c$ *in* C *is canonical model for a presentation* P *on language* $\mathcal{L}$, *iff the assignment* $c \to a$ *maps* C *onto* M, *i.e.* $M = (a\colon c \in C)$.

Generic diagrams allow us to define canonical models with specific functions. The CTA's are specific canonical models. In the sections above we have shown how to define models with G-diagrams. Initial models are models definable by G-diagrams. Let us carry the initial models onto the planning foundations.

THEOREM 4.2. *For the free proof trees defined for a goal formula from the* GF*-diagram there is an initial model satisfying the goal formulas. It is the initial model definable by the* GF*-diagram.*

PROOF. In planning with GF-diagrams plan trees involving free Skolemized trees is carried along with the proof tree for a plan goal. The idea is that if the free proof tree is constructed then the plan has a model in which the goals are satisfied. There is analogy to SLD proofs. We can view on the one hand, SLD resolution type proofs on ground terms, where we go from $p(0)$ to $p(f(c))$; or form $p(f(c))$ to $p(f(g(c)))$. Whereas, while doing proofs with free Skolemized trees we are facing proofs of the form $p(g(\ldots))$ proves $p(f(g(\ldots)))$ and generalizations to $p(f(x))$ proves for all x, $p(f(x))$. Since the proof trees are either proving plan goals for formulas defined on the GF-diagram, or are computing with Skolem functions defining the GF-diagram, the model defined by the GF-diagram applies and it is initial for the proofs. □

For free Skolemization what comes to our logician mind's side of things is Generic model expansion. It is relevant to our methods of modeling, planning, and reasoning. A free proof tree is a proof tree that is constructed with free Skolem functions and GFS from a GF-diagram. The idea is that if the free proof tree is constructed for a goal formula, the GF-diagram defines a model satisfying the goal formula satisfied. The model is the initial model of the AI world for which the free Skolemized trees were constructed. Thus we have transformed the model theoretic problems of computing to that of defining computing with Generalized Diagrams. There are recent papers since 1992 (Goguen 1973, Nourani and Hoppe 1994), where we have presented this theory and technique. Partial deductions in the present approach correspond to proof trees that have free Skolemized trees in their representation. These concepts will be developed in forthcoming papers of this author. In the present approach, the free proof tree technique, as we shall further define, leaves could be virtual, where virtual leaves are free Skolemized trees.

5. GENERALIZED PROOFS AND HILBERT MODELS

For plans with free Skolemized trees we can apply the Hilbert epsilon technique to define computing models. What applying Hilbert's epsilon implies is that there is a model M for the set of formulas such that we can take an existentially quantified formula $w[X]$ and have it instantiated by a Skolem function that can answer the membership question to the model. Whether or not Hilbert had intended it this way or not is relevant at present. The issue in our approach, however, is that we are not so much concerned with existentially quantified formulas. We start with some Skolem functions to define initial models.

Thus we have Hilbert models for which the Skolem functions implicitly define membership to the set defining a model. Thus we can refer to such models in our approach as Hilbert models. Now to address the issues of proof trees in connection with the formulation in Nourani and Hoppe (1994). We have planning applications in which there is goal formula to be satisfied with perhaps existential quantifiers. Since we are interested in model theoretic techniques for handling proofs with the method of free proof trees we propose the following model-theoretic view, which we refer to by the Hilbert model theorem for Skolemized algebraic tree computing.

THEOREM 5.3. *The Hilbert's epsilon technique implies there is a model* M *for the set of formulas such that we can take an existentially quantified formula* $w[X]$ *and have it instantiated by a Skolem function which can answer the satisfiability question for the model.*

PROOF. We start with some Skolem functions to define the models. The equivalence defined by Hilbert's epsilon function is instantiated by the Skolem functions defining GF-diagrams. Thus we have Hilbert models for which the Skolem functions implicitly define membership to the set defining a model. While doing proofs with free Skolemized trees we are facing proofs of the form $p(g(\ldots))$ proves $p(f(g(\ldots)))$ and generalizations to $p(f(x))$ proves for all x, $p(f(x))$. Since the proof trees are either proving plan goals for formulas defined on the G-diagram, or are computing with Skolem functions defining the GF-diagram, the model defined by the GF-diagram applies for the proofs. □

6. ORDINAL PLANNING WITH GF-DIAGRAMS

An important application area of the present formulation of reasoning is in robot planning (Fikes and Nilsson 1971). The significance of plausibility reasoning in this context is its ability to make minimal changes to a belief state when given new information. Suppose we represent a robot's beliefs by a set of propositions, which are a set of subsets of W. If the robot already believes a proposition p, then the new information does not affect its belief state. If it does not believe p, then the impact of the new information on the robot's belief state has to be defined. The revision problem is handled by partitioning the possible worlds and ordinal conditional functions (OCF) as in Nourani (1991). The revision system can be used to change the belief state of the configuration of the blocks in the blocks world problem solver. For example, if one tells the robot to put block C on top of block A, the truth-sets are changed to have this new belief. The revision system will automatically revise its belief about the configuration of blocks so as to accommodate the new piece of information.

Furthermore, it will revise its beliefs in a minimal manner. Use of nondeterminism by GF-diagrams is useful in realizing robot plans since there usually are many indeterminate formulas in the world diagram, whose actual values are inconsequential in an immediate plan of action, thus may be assigned a free Skolem function to satisfy a goal. For example, it does not matter whether block D is free or not, if the goal is to have A on B, the GF assignment to free(D) is F_free(D), a Skolem function. Such assignment on diagrams allows planning to define relations in terms of known relations and to implement free proof systems for goal satisfiability by defining equivalence relations on terms based on Skolem functions such as F_free(t) for a term t. The theorem proving implications of the above and planning with free proof trees are to be a subject of forthcoming research papers of the present project. This approach provides a solution to what is known as the "frame problem" (Nourani 1991) in robot planning. Furthermore, by the information updates to the plausible diagram one can perform model reconstruction to implement dynamic replanning in which the world model is constantly changed and updated. Planning and goal satisfiability goes on by a proof system that checks satisfiability of formulas by term reduction to free Skolem terms defining GF diagrams.

7. GF PLANS AND GENERALIZED FREE PROOFS

7.1. GF Plans and Proof Abstraction

We propose methods that can be applied to planning with GF-diagrams such that similar effects to partial deductions (Hoppe 1994) can be achieved. In planning with GF-diagrams that part of the plan that involves virtual Skolemized trees is carried along with the proof tree for a plan goal (e.g., see Gen-Nils 1987). The idea is that if the free proof tree is constructed then the plan has a model in which the goals are satisfied (Nourani 1995a). The planning process at each stage can make use of GF-diagrams by taking the free interpretation, as tree-rewrite computations as in Artificial Intelligence (1980), for example, of the possible proof trees that correspond to each goal satisfiability. Planning with GF-diagrams carries along that part of the plan that involves free Skolemized trees, with the proof tree for a plan goal. The idea is that if the free proof tree is constructed then the plan has a model in which the goals are satisfied. The model is the initial model of the AI world for which the free Skolemized trees were constructed. Partial deductions in this approach correspond to proof trees that have free Skolemized trees in their representation. The analogy is SLD resolution type proofs on ground terms, where we go from $p(0)$ to $p(f(c))$; or form $p(f(c))$ to $p(f(g(c)))$. Whereas, while doing proofs with free Skolemized trees we are facing proofs of the form $p(g(\ldots))$ proves $p(f(g(\ldots)))$ and generalizations to $p(f(x))$ proves for all x, $p(f(x))$. Thus the free proofs are in some sense an abstract alternative to SLD proofs.

7.2. Proofs Trees and Hilbert Models

What Hilbert's epsilon approach implies is that there is a model M for the set of formulas such that we can take an existentially quantified formula $w[X]$ and have it instantiated by a Skolem function that can answer the membership question to the model. Whether Hilbert had intended it this way is not relevant. The issue in our approach, however, is that we are not so much concerned with existentially quantified formulas. We start with some Skolem functions to define Initial models. Thus we have Hilbert models for which the Skolem functions implicitly define membership to the set defining a model. For example can refer to Nourani (1980) for definition of epsilon models of set theory. Thus we can refer to such models in our approach as Hilbert models. How to address the issues of proof trees in connection with the formulation in Poole (1987). We have planning applications in which there is goal formula to be satisfied, rather than explained, from a set of premises that are facts and defaults. Therefore, we have designed an algorithm applicable to proof abstraction, planning and EBG. The computation method for goal satisfiability allowing ground instances of defaults (those without variables and Skolem constants). We present the following algorithm for planning with free proof trees.

Skolemized Virtual Proof Tree Planning Algorithm

1. Skolemize F, the set of facts, forming F_{sk} (free variables are universally quantified)
2. Try to prove g, from the elements of F_{sk} and defaults DF as axioms. Let DF0 be the instances of default used in the proof.
3. Form DF1 by replacing free variables in DF0 with unique constants.
4. Add DF1 to F and try to prove an inconsistency. If a complete search for a proof fails, the goal of the plan is satisfied.

The algorithm in what (Poole 1987) can be generalized and applied with GF-trees to planning and FTP with virtual proof trees. Since the proof methods can make use of free tree congruences (Nourani 1995a), step 4 of the above algorithm can be implemented without any real difficulty. In step (3) above it might be possible to replace free variables with elements of F_sk, and at times might make free proof more efficient. These are topics for future research.

7.3. Recursion and Robots

The CM construction (Franova 1995) transforms the problem of proving an atomic formula to solving equations, also could lend itself to the formulations above. The formulation in my papers are applicable in formulating a natural decomposition for CM and Recursion (Fran-Kooli 1997). The language we start with are has explicit type construction in its syntax. That is why it is easy to view the planning proof problems as that of an inductive decomposition (Franova 1995) and abstract proof trees which define models where a goal is satisfied. It is a way for us to start applying CM to model-theoretic planning. The ideas are close counterparts since model-theoretic planning and the planning algorithm, Section 4.2, are constructible models for plan goals. The planning techniques in our papers are ways to get to goals by constructing models Skolem tree by Skolem tree. Combining Section 3 where block diagrams are defined with Section 7 block computation examples as in Fran-Kooli (1997) can be carried out.

The How to Clear a Block problem, or HCB, is stated by the formula $\forall s_0 : b \exists z_1$. Clear $(z_1, s_0 : b)$, there is a state z_1 for object called b in state s_0, such that b becomes clear. Starting with a specified table configuration the set of facts are Skolemized with terms defined as Section 3 "put" function terms $z = \text{put}(x, y, \text{Tab})$, or $z = \text{put}(x, w, \text{put}(w, y, \text{Tab}))$. The goal formula is satisfied by the formula sequences generated by the Skolemized virtual proof tree planning algorithm.

8. VISUAL PLANNING

8.1. Morph Gentzen VR

The IM Morphed computing logic for computing for multimedia is new projects with important computing applications since Nourani (1927, 1996a). The basic principles are a mathematical logic where a Gentzen (1943) or natural deduction (Prawitz 1972) systems is defined by taking arbitrary structures and multimedia objects coded by diagram functions. Multimedia objects are viewed as syntactic objects defined by functions, to which the deductive system is applied. Thus we define a syntactic morphing to be a technique by which multimedia objects and hybrid pictures are holographically mapped via their defining functions to a new hybrid picture. Functorial topological structures can be defined without difficulty. The deduction rules are a Gentzen system augmented by Morphing, and transmorphing.

The logical language has function names for hybrid pictures. The *MIM Morph Rule*—An object defined by the functional n-tuple $\langle f_1, \ldots, f_n \rangle$ can be Morphed to an object defined by the functional n-tuple $\langle h(f_1), \ldots, h(f_n) \rangle$, provided h is a homorphism of abstract signature structures (Nourani 1993c). The *MIM TransMorph Rules*—A set of rules whereby combining hybrid pictures $p_1, \ldots, p_n$ defines an event $\{p_1, p_2, \ldots, p_n\}$ with a consequent hybrid picture p. Thus the combination is an impetus event. By transmorphing hybrid picture's corresponding functions a new hybrid picture is deduced. The techniques can be applied to arbitrary topological structures. The languages and MIM rules are applied to algebraic structures. The deductive theory is a Gentzen system in which hybrid pictures are named by parameterized functions; augmented by the MIM Morph and transform rules. The model theory is defined from intelligent syntax languages (Nourani 1995d, 1996a). A computational logic for intelligent languages is presented in brief with a soundness and completeness theorem in Nourani (1996a). The idea is to do it at abstract models syntax trees without specifics for the shapes and topologies applied. We start with $\mathcal{L}_{\omega 1,\omega}$ and further on might apply well-behaved infinitary languages. A soundness and completeness theorem has been put forth (Nourani 1997) stating the MIM soundness and completeness:

THEOREM 8.4. *Nourani* (1997) *MIM Logic is sound and complete.*

8.2. Intelligent Trees and Spatial KR

Visual objects connected by agents carrying information represent the visual field amongst objects about the field, and carried onto intelligent trees for computation. Intelligent trees compute the spatial field information with the diagram functions. The trees defined have function names corresponding to computing agents. The computing agent functions have a specified module defining their functionality. By an intelligent language we intend a language with syntactic constructs that allow function symbols and corresponding objects, such that the function symbols are implemented by computing agents. A set of function symbols in the language, referred to by AF, is the set modeled in the computing world by AI agents with across and/or over board capability.

8.3. Spatial Information and Morph Gentzen

Multiagent spatial vision techniques are introduced in the paper. The duality for our problem solving paradigm (Nourani 1991, 1993c) is generalized to be symmetric by the present paper to formulate double vision computing. The basic technique is that of viewing the world as many possible worlds with agents at each world that compliment one another in problem solving by cooperating. An asymmetric view of the application of this computing paradigm was presented by the author and the basic techniques were proposed for various AI systems (Nourani 1991). The double vision computing paradigm with objects and agents might be depicted by the following figure. For computer vision (Winston 1975), the duality has obvious anthropomorphic parallels. The object coobject pairs and agents solve problems on boards by cooperating agents. The cooperative problem solving paradigms have been applied ever since the AI methods put forth by Hays-Roth (1985). However, the multiagent multiboard techniques due to Nourani (1995b) planning visual spacecraft navigation by multiagent multimedia are applications (Nourani 1999a).

8.4. Visual Context and Objects

The visual field is represented by visual objects connected with agents carrying information amongst objects about the field, and carried onto intelligent trees for computation. Intelligent trees compute the spatial field information with the diagram functions. The trees defined have function names corresponding to computing agents. The computing agent functions have a specified module defining their functionality. The balloons are visual objects, the squares agents; the dotted lines the message paths. Multiagent spatial vision techniques are introduced in Nourani (1998a, b). The duality for our problem solving paradigm (Nourani 1991, 1995a, b) is generalized to be symmetric by the present paper to formulate Double Vision Computing. The basic technique is that of viewing the world as many possible worlds with agents at each world that compliment one another in problem solving by co-operating. The author presented an asymmetric view of the application of this computing paradigm and the basic techniques were proposed for various AI systems (Nourani 1991). The double vision computing paradigm with objects and agents might be depicted by the following figure. For computer vision (Winston 1975), the duality has obvious anthropomorphic parallels. The object co-object pairs and agents solve problems on boards by co-operating agents.

8.5. Multiagent Visual Planning

The co-operative problem solving paradigms have been applied ever since the AI methods put forth by Hays-Roth et al. (1985). See Nii (1986). The multiagent multi-board techniques due to Nourani (1995a), see next section. The BID (Brazier-Treur et al. 1997) model has to be enhanced to be applicable to intelligent multimedia (Nourani 1998c, 1999a). Let us start with an example multi-board model where there multiagent computations based on many boards, where the boards corresponds to either virtual possible worlds or to alternate visual views to the world, or to the knowledge and active databases. The board notion is a generalization of the blackboard problem solving model (Hays-Roth 1985, Nii 1986).

The blackboard model consists of a global database called the blackboard and logically independent sources of knowledge called the knowledge sources. The knowledge sources respond opportunistically to the changes on the blackboard. Starting with a problem the blackboard model provides enough guidelines for sketching a solution. Agents can cooperate on a board with very specific engagement rules not to tangle the board nor the agents.

8.6. Model Functions and Visual Objects

The functions applied to define models via G-diagrams which are corresponded in design with selectors to multiagent visual objects. The following figure is an overview to the design.[1]

Agents are squares and objects the balloons. The dotted lines are message paths. The balloons are objects applied to compute field information with agents where all computation is expressed and carried on intelligent tree language (Nourani 1996a). Every function $F(x_1, \ldots, x_n)$ might correspond to visual multiagent objects, as in the above figure, where the x_i's might be instantiated with agent variables and selectors to virtual or real objects. A visual world model is hence defined to plan with.

[1] The planning and diagram function coding techniques are the property of the author's venture business. Commerical and R&D applications must be with proper credit and permission.

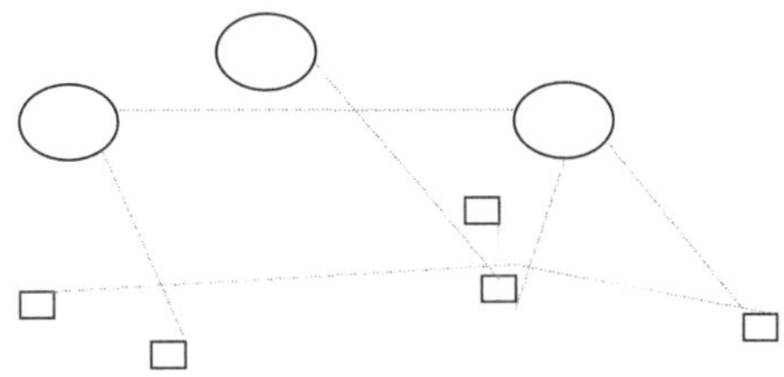

Figure 1. Multiagent objects.

9. CONCLUSIONS

The paper has presented techniques for planning and automated reasoning with free proof trees. Generalized diagrams are applied to define models and design free proof trees. Free proofs are based on free Skolemization of diagrams. A model theoretic view to abductive computing is defined by introducing predictive diagrams. By defining predictive diagrams partial deduction and abduction could be represented model-theoretically (Nourani 1999b). Hilbert models are defined for free proof trees to put forth a basis for a model theory with free proofs for AI applications. Initial models are presented for diagram computations. An algorithm is presented for automated plan goal proofs and plan computation with virtual proof trees. Applications to robot planning is outlined. A specific robot planning example with recursion is briefed. Visual multiagent planning is presented in brief based on the model plans developed. GF plans and proof abstraction applications are alluded to and a Skolemized virtual proof tree planning algorithm is presented. Visual planning is new applications with Morph Gentzen VR, intelligent trees and spatial KR. Applications to model discovery is briefed in Nourani (1999a).

EXERCISES

1. a. What is the plan goal satisfaction problem when the proof trees are agent virtual trees?
   b. Design a two-agent planner that can allow agents a1 and a2 play a virtual volleyball game on a variable height net. State every simplifying assumption.
2. The concept formation problem is a common type of inductive inference problem. The data assert a common property of some objects and deny that property to the others. The inductive hypothesis is a universally quantified sentence that summarizes the conditions under which an object has that property.
   2a. Show that agent signatures can be applied to assert properties to specific agents and define concept formation.
   2b*. Apply the algorithm in Section 7.2 to devise virtual concept formation towards inductive inference. State what simplifying assumptions are necessary to deploy your design.

3*. The generalization inference theorem (c.f. Genesereth-Nilsson) that provability can be inferred on a universally quantified formulas if the variable quantified does not occur free on a provable formula, has a counterpart onto GF virtual proof trees. Can you state what the counterpart might be.

4. Based on the last paragraph Section 7.3. The How to Clear a Block problem, or HCB, is stated by the formula $\forall s_0 : b \exists z_1$. Clear $(z_1, s_0 : b)$, there is a state z_1 for object called b in state s_0, such that b becomes clear. Starting with a specified table configuration the set of facts are Skolemized with terms defined as Section 3 "put" function terms $z = \text{put}(x, y, \text{Tab})$, or $z = \text{put}(x, w, \text{put}(w, y, \text{Tab}))$. Show that the goal formula is satisfied by the formula sequences generated by the Skolemized virtual proof tree planning algorithm.

* Nontrivial.

REFERENCES

1. F. M. T. Brazier, B. Dunin-Keplicz, N. R. Jennings, and J. Treur, "DESIRE: Modelling Multi-Agent Systems in a Compositional Formal Framework" (M. Huhns and M. Singh, Eds.), Vol. 1. International Journal of Cooperative Information Systems, 1997.
2. R. E. Fikes and N. J. Nilsson, *Artificial Intelligence* 2, 189 (1971).
3. M. Fraonova, "CM Strategy: A Methodology for Inductive Theorem Proving or Constructive Well-Generalized Proofs" (A. Johsi, Ed.), pp. 1214–1220. Proc. Ninth IJCAI, Los Angeles, 1995.
4. M. Franova and M. Kooli, "Recursion Manipulation for Robotics: Why and How?" Proceedings EMCSR98, Universite Paris Sud, Orsay, Vienna, Austria, 1997.
5. J. Galsgow et al., "Diagrammatic Reasoning—Cognitive and Computational Perspectives." (B. Chandrasekaran and N. H. Narayaran, Eds.), AAAI Press, 1996.
6. M. Genesereth and N. J. Nilsson, "Logical Foundations of Artificial Intelligence," Morgan Kaufmann, 1987.
7. G. Gentzen, *Math. Ann.* 119, 140 (1943).
8. J. A. Goguen, J. W. Thatcher, E. G. Wagner, and J. B. Wright, "A Junction Between Computer Science and Category Theory." IBM Research Report RC4526, 1973.
9. J. A. Goguen, J. W. Thatcher, E. G. Wagner, and J. B. Wright, "An Algebraic Approach to Abstract Data Type Specifications and Their Correctness of Implementations" (R. Yeh, Ed.), Vol. IV. Current Trends in Programming Methodology, Data Structuring Techniques. Prentice Hall, 1978.
10. B. Hays-Roth, *Artificial Intelligence* 26, 251 (1985).
11. Th. Hoppe, "On the Relationship Between Partial Deduction and Abductive Reasoning," pp. 154–158. Proc. 10th ECAI Conference. John Wiley and Sons, Vienna, Austria, 1994.
12. I. Mosetic and C. Holsbaur, "Extending Explanation Based Generalization by Abstraction Operators," Vol. 482. Machine Learning EWSL-91, Springer-Verlag, LNAI, 1991.
13. J. P. Muller and M. Pischel, "Modeling Interactive Agents in Dynamic Environments," Proceedings 11th European Conference on AI. John Wiley and Sons, Amsterdam, The Netherlands, 1994.
14. P. H. Nii, *The AI Magazine* 38, (1986).
15. C. F. Nourani, "A Model-Theoretic Approach to Specification, Extension, and Implementation of Data Types," Proc. International Symposium on Programming, Paris. Springer-Verlag, Berlin, 1980.
16. C. F. Nourani, "Equational Intensity, Initial Models, and AI Reasoning" (Tim O'Shea, Ed.). Technical Report: A Conceptual Overview, in Proc. Sixth European Conference in Artificial Intelligence. Pisa, Italy. North-Holland, 1984.
17. C. F. Nourani, "Diagrams, Possible Worlds, and the Problem of Reasoning in Artificial Intelligence," European Summer Logic Colloquium, Padova, Italy, 1987.
18. C. F. Nourani, "Diagrams, Possible Worlds, and the Problem of Reasoning in Artificial Intelligence." Proc. Summer Logic Colloquium, Padova, Italy, 1988.
19. C. F. Nourani, "Planning and Plausible Reasoning in AI," pp. 150–157. Proc. Scandinavian Conference in AI, IOS Press, Roskilde, Denmark, 1991.
20. C. F. Nourani, "Free Proof Trees and Model-Theoretic Planning." AISB, Sheffield, England, 1995a.
21. C. F. Nourani, "Double Vision Computing." IAS-4, Intelligent Autonomous Systems, Karlsruhe, Germany, 1995b.
22. C. F. Nourani, "Visual Computational Linguistics—The Preliminary Definition." 34th International Linguistics Colloquium, Mainz, Germany, 1995c.
23. C. F. Nourani, "MIM Logik." Summer Logic Colloquium, Prague, 1998a.
24. C. F. Nourani, "A Parallel Functional Modular Language with Agents and Distributed Objects." Brief, IFL Wkshp Parallel Functional Languages, London, 1998b.
25. C. F. Nourani, "Intelligent Multimedia," Intelligence and Multimedia Applications. Poster Announcement, Monash University, Victoria, Australia, 1998c.
26. C. F. Nourani, "Intelligent Trees, Thought Models, and Intelligent Discovery." Model-Based Reasoning in Scientific Discovery, Pavia, Italy, 1998d.
27. C. F. Nourani, "Intelligent Multimedia—New Computing Techniques and Its Applications." CSIT, Moscow, 1998e.
28. C. F. Nourani, "Double Vision Spacecraft and Autonomous Flight Control." AA99 Autonomy Control Track, Seatlle, WA, 1999a.
29. C. F. Nourani, "Nonmontonic Model Culmination." Proceedings ASL, UCSD, La Jolla, 1999b.
30. C. F. Nourani and Th. Hoppe, "GF Diagrams for Models and Free Proof Trees," Berlin Logic Colloquium, Germany, 1994.
31. C, F. Nourani and Th. Hoppe, "G-Diagrams for Models and Free Proof Trees, and Abductive Automated Reasoning," Proc. Automated Reasoning-AISB. England, 1995.
32. D. Poole, "Variables in Hypotheses." Proc. IJCAI 87, Milan, Italy, 1987.
33. D. Prawitz, "Natural Deduction: A Proof Theoretic Study" Almqvist and Wiksell, Stokhlom, 1973.
34. E. Sacerdti, *Artificial Intelligence* 5, 115 (1974).
35. Special Issue on Non-monotonic Logic, *Artificial Intelligence*, (1980).
36. W. Sphon, "Ordinal Conditional Functions: A Dynamic Theory of Epistemic States" (W. L. Harper and B. Skyrms, Eds.), pp. 105–134. Causation, in Decision, Belief Change, and Statistics, Kluwer Academic Publishers, 1988.
37. D. Wilkins, Domain Independent Planning: Representation and Plan Generation, *Artificial Intelligence* 22, 269 (1984).

CHAPTER 14

Intelligent Cyberspace Interfaces and Wireless Access Protocol Generic Computing

CONTENTS

1. INTRODUCTION

The application of intelligent mobile objects to business computing is based at various diverse sectors of industry on design properties, behavior, relationships, and engagement rules. Intelligent authenticated interfaces to WWW applications apply the new intelligent cybersignatured computing techniques to business objects and EC applications. The paper defines the areas and presents WWW authentication principles. Cybersignatures and automatic authentication can be designed and computed with agent language processing.

The term "agent" refers to the AI constructs that enable computation on behalf of an AI activity, for example Genesereth and Nilsson (1987), and Brazier et al. (1995). It also refers to computations that take place in an autonomous and continuous fashion, while considered a high-level activity, in the sense that its definition is software and hardware implementation, independent. *Software agents* are specific agents designed by a language that carry out

ISBN: 1-58883-037-3/$35.00

Intelligent Multimedia Computing Science
By Cyrus F. Nourani
Pages: 177–186

specified tasks and define software functionality. Most agents defined by our examples are software agents. Objects are in the well-known sense of the word in object programming, abbreviated by OOP. Objects consist of abstract data, perhaps encapsulation, and operations. Most recent programming techniques apply OOP. IOOP (Nourani 1998d) is a recent technique developed by the author combining AI and software agents with OOP. Other application areas defined are object modules programming paradigms for composing modules. KB is an abbreviation for knowledge bases. The separation makes large applications more tractable, global analysis feasible, and software reusable. Modular languages deal with languages, techniques, and tools for the design and implementation of large-scale software systems in a modular and extensible way. Languages, techniques, and tools for the design and implementation of large-scale software systems in a modular and extensible way. JAVA, Modula, and SAP, are few examples.

For our project the modular programming concepts are combined with software agent computing, new intelligent OOP-IOOP object-coobject pairs. CORBA views designs with objects to be either business objects, A business process object, or a presentation object. For these applications, an *object* is defined to be a uniquely identifiable real-world entity. The wireless world has brought forth new computing paradigms and languages, for example, J2ME, to minimize computation onto wireless mobile interfaces. The computing over the wireless networks is via WAP. Examples are Alcatel WAP. Certicom Cryptography with WAP services, WIWA: Windows tool for generating WAP portals with one or more content providers or personal WML homepages. It allows fast and simple WAP sites construction without any programming skills, WAP or WML knowledge. Ericsson's WAP developer's zone information and tools to build applications for Ericsson's WAP devices MobileDev is the first. MobileDev supports WAP technologies like WML, HDML, Microsoft Active Server Pages (ASP), Perl and Java Server Pages (JSP). AVIDWireless Rapid Tools (ART) is a Java component that allows development of WAP/WML, i-Mode/MS Mobile Explorer/CHTML, Palm VII/PQA and HTML wireless applications.

The questions addressed are: Starting with Crypto Mobile Functions
Wireless Device Server Processing Network Multicontrol Crypto Assets
Networking Wireless Assets

Our goal is to provide a computing paradigm generic to accommodate mobile wireless. Computations without having to reinvent the basic network cybercomputing systems and the safety and privacy bases.

2. HETEROGENEOUS COMPUTING AND WIRELESS ACCESS PROTOCOL

Modular languages deal with languages, techniques, and tools for the design and implementation of large-scale software systems in a modular and extensible way. CORBA, JAVA, and SAP are specific examples. These languages do not have specific agent computing support. For our project the modular programming concepts are combined with software agent computing, new IOOP constructs object-coobject pairs and kernels. Modules are aggregate objects with a specific functionality defined. Parallel module specification what we intend by a parallel specification language is one that allows a user to define parallel modules with embedded communicating processes. The user supplies the type of parallelism and the types of message processes are expected to communicate. The processes are embedded within a module, and there is syntax provided for communicating pairs of processes.

2.1. Agent Language Processing and Models

By an intelligent language, we intend a language with syntactic constructs that allow function symbols and corresponding objects, such that the function symbols are implemented by computing agents. Sentential logic is the standard formal language applied when defining basic models. The language $\mathscr{L}$ is a set of sentence symbol closed by finite application of negation and conjunction to sentence symbols. Once quantifier logical symbols are added to the language, the language of first order logic can be defined. A Model $\mathscr{A}$ for $\mathscr{L}$ is a structure with a set A. There are structures defined for $\mathscr{L}$ such that for each constant symbol in the language there corresponds a constant in A.

For each function symbol in the language there is a function defined on A; and for each relation symbol in the language there is a relation defined on A. For the algebraic theories we are defining for intelligent tree computing in the forthcoming sections the language is defined from signatures as in the logical language is the language of many-sorted equational logic. The signature defines the language $\mathscr{L}$ by specifying the function symbols' arities. The model is a structure defined on a many-sorted algebra consisting of S-indexed sets for S a set of sorts. A set of function symbols in the language, referred to by AF, is the set modeled in the computing world by AI agents with across and/or over board capability. Thus, the language $\mathscr{L}$ defined by the signature has designated function symbols called AF. For the specific foundations we refer to Nourani (1998b), for example. The AF function symbols define signatures that have specific message paths defined for carrying context around an otherwise context free abstract syntax. A set of function symbols in the language, referred to by AF, is the set of agents with nontrivial capability. An abstract language that is capable of specifying modules, agents, and their communications expresses the computation. We have put together the AI concepts with syntactic constructs that could run on the tree computing theories we are presenting in brief. Intelligent languages and signatures allow us to present computational theories with formulas on terms with intelligent function symbols. The project has further led to foundations to computing with intelligent trees. The techniques for generating intelligent models are developed with soundness and completeness theorems (Nourani 1994, 1998b). The models can be applied as a basis to authentication on cyberspace computing and WWW computing logics.

2.2. Authentication and the World Wide Web

A basis for agent computing with intelligent languages and intelligent trees is presented with applications to WWW interfaces. We present intelligent syntax and put forth intelligent tree computing. Multiagent signatures are defined and applied to define the basis for tree information-theoretic computing and agent cyberspace applications. The project is applicable to design multiagent protocol and has been applied to put forth cryptosignatures. The project has further led to foundations to computing with intelligent trees. Intelligent game trees are defined with chess playing examples and applications to cyberspace computing. Techniques for generating intelligent models are developed with soundness and completeness theorems abbreviated here with basis in our papers. The models can be applied as a basis to authentication on cyberspace computing. The WWW applications are summed by an authentication proposition. Nourani (1998c) has applied agent computing to design business co-operating systems via the cyberspace.

Our computing by agents and multiboards apply cooperative problem solving techniques on the individual boards, but with generalized paradigms encompassing multiagents cross the boards. Privacy and security for AI systems designed by multiagent techniques is defined in terms of agents, functions, objects, and kernels. Obligations and sharing resource necessities and norms are defined. It is shown how the methods could be superimposed onto the existing theories for privacy and security of computer systems.

A brief theoretical conclusion on soundness of our formulation is stated to complement the practical methods put forth. This paper is based on a startup paper for the on multiagent privacy and security of AI systems written buy the author in 1993. The following sections develop the techniques and theory of computing with trees on signatures that bear agent functions on trees. Our results for computability of initial models by tree rewriting are developed further for as a foundation for computing on trees to be applicable to intelligent tree computing (Nourani 1994). The concepts of intelligent syntax and intelligent languages are presented. The present applications are tree computing, agent authentication, intelligent WWW interfaces, and multiagent protocol. Tree rewriting is defined on intelligent trees by presenting the concepts and definition of algebraic tree intelligence content and mutual tree intelligence content within a forest. At the forest suddenly a tree information theoretic theorem had presented itself, defining a correspondence between intelligent tree rewriting and tree intelligence preservation.

Models for intelligent theories emerge from the algebraic intelligent tree rewriting. Intelligent algebraic tree completion theorems and initial model rewrite theorems are put forth for intelligent trees. A soundness and completeness theorem is proved for intelligent tree rewriting as a formal model basis. A proposition sums the applications to WWW. The first author and others had suggested intelligent agents could be applied to Web Browsers since 1994. An article in Computer World, January 1997, features "WBI" and quotes "the most productive internet tools you don't waste time looking for things that you don't need." It has been claimed agents can help us at the sea of information that grows and surrounds us everyday.

2.3. Agent Language Processing

By an intelligent language we intend a language with syntactic constructs that allow function symbols and corresponding objects, such that the function symbols are implemented by computing agents. Sentential logic is the standard formal language applied when defining basic models. The language $\mathcal{L}$ is a set of sentence symbol closed by finite application of negation and conjunction to sentence symbols. Once quantifier logical symbols are added to the language, the language of first order logic can be defined. A model $\mathcal{A}$ for $\mathcal{L}$ is a structure with a set A. There are structures defined for $\mathcal{L}$ such that for each constant symbol in the language there corresponds a constant in A. For each function symbol in the language there is a function defined on A; and for each relation symbol in the language there is a relation defined on A. The signature defines the language $\mathcal{L}$ by specifying the function symbols' arities.

A set of function symbols in the language, referred to by AF, is the set modeled in the computing world by AI agents with across and/or over board capability. Thus the language $\mathcal{L}$ defined by the signature has designated function symbols called AF. For the specific foundations we refer to Nourani (1998b), for example. The AF function symbols define signatures that have specific message paths defined for carrying context around an otherwise context free abstract syntax. A set of function symbols in the language, referred to by AF, is the set of agents with nontrivial capability. An abstract language that is capable of specifying modules, agents, and their communications expresses the computation. Intelligent languages and signatures allow us to present computational theories with formulas on terms with intelligent function symbols. Computing with intelligent trees might embed automatic agent signature authentication. The basis for the computing is a signature match to start agent messages. Basic authentication is automatic when the intelligent computing interface to WWW is cranked up. Further encoding can be embedded when the agent signatures are designed. For example, the agent signature can specify what the agent types are with which it can engage exchange at all. The signature might specify the class of messages and or transactions allowed. The agent signatures can also encode basic agent protocol. The Cyberdocking project Nourani (1997) and Section 6 has specific modular docking facilities at the cyberspace.

2.4. Intelligent Trees and the WWW

When it comes to intelligent interfaces to the WWW the basic applications for intelligent game trees is in managing to acquire and complete a critical business cooperative transaction at minimal time when there are agents competing on the WWW. Intelligent game tree strategies are not always based on ordinary game principles. Without further investment, it is not prudent to elaborate on the details here. Starting with basic authentication as presented in Section 2.3, where authentication is an abstract term referring to matching signatures to some specified signatures, with authentication encoded onto the signatures. *The Authentication Proposition* WWW intelligent interface computing incorporates basic agent authentication via intelligent tree signatures (Nourani and Moudi 2000). At the present stage, it might sound childish to think agents play games at WWW, but AI computing is on its course onto agent computing and marketing via virtual communities (Riggins and Sue 1998, Hagel and Armstrong 1997).

3. MOBILE AGENTS AND JAVA

Mobile agents are agents that can physically traverse across a network, and perform tasks on agent hosting machines. The mobility allows processes to move from computer to computer, for processes to split into multiple instances that execute on different machines, and to return to their point of origin. The process movements allow executable code to travel and interact with databases, file systems, information services and other agents. Aglets (Nourani 1997), an innovation developed by IBM Japan are specific mobile agent carriers. An aglet is a Java-based autonomous software agent. The Java aglet extends the model of network-mobile code made famous by Java applets. Like an applet, the class files for an aglet can migrate across a network. But unlike applets, when an aglet migrates it also carries its state.

An applet is code that can move across a network from a server to a client. An aglet is a running Java program (code and state) that can move from one host to another on a network. In addition, because an aglet carries its state wherever it goes, it can travel sequentially to many destinations on a network, including eventually returning back to its original host. The following things are required to allow agents to migrate across a network

1. Common execution language
2. Process persistence
3. Communication mechanism between agent hosts
4. Security to protect agents and agent hosts

An aglet can experience many events in its life. It can be:

Created: a brand new aglet is born—its state is initialized, its main thread starts executing
Cloned: a twin aglet is born—the current state of the original is duplicated in the clone
Dispatched: an aglet travels to a new host—the state goes with it
Retracted: an aglet, previously dispatched, is brought back from a remote host—its state comes back with it
Deactivated: an aglet is put to sleep—its state is stored on a disk somewhere
Activated: a deactivated aglet is brought back to life—its state is restored from disk
Disposed of: an aglet dies—its state is lost forever

The state of the execution stacks and program counters of the threads owned by the aglet are not serialized. Object serialization affects only data on the heap, not the stacks or the program counters. Thus when an aglet is dispatched, cloned, or deactivated, any relevant state sitting on any stack of a running aglet, as well as the current program counter for any thread, is lost. Before it is serialized, an aglet must place on the heap everything it will need to know to be brought to life properly as a newly activated aglet, a freshly dispatched aglet, or a clone. It can't leave any of this information on the stack, since the stacks are not reproduced in the aglet's new instantiation. As a result, the aglet host informs an aglet that it is about to be serialized so that the aglet can prepare itself. When the aglet is informed of an impending serialization, it must place onto the heap any information it will need to continue its execution properly when it is brought up. From the practical stand aglets are finite state machine with the heap as the sole repository of the machine's state. If at any point in an aglet's life you can know what state it is in by looking at its heap, then it can be serialized at any time. If not, then you must have a way to record sufficient information on the heap just before serialization.

3.1. Mobile Agents and WWW

A Java aglet is similar to an applet in that it runs as a thread (or multiple threads) inside the context of a host Java application. To run applets, a Web browser fires off a Java application to host any applets it may encounter as the user browses from page to page. That application installs a security manager to enforce restrictions on the activities of any untrusted applets.

To download an applet's class files, the application creates class loaders that know how to request class files from an HTTP server. An aglet requires a host Java application to be running on a computer before it can visit that computer. When aglets travel across a network, they migrate from one aglet host to another. Each aglet host installs a security manager to enforce restrictions on the activities of untrusted aglets. Hosts upload aglets through class loaders that know how to retrieve the class files and state of an aglet from a remote aglet host.

Cybersafety and Aglet Life: A necessary condition to attain cybersafety with mobile agent computing with aglets is that the Aglet life style be sound with respect to certain basic design goals stated in Nourani and Moudi (2001).

3.2. Cyberspace Safety

With the advancement of multiagent AI techniques and their increasing applications to real problems of computing basic sound design and operations principles are critical necessities. Cyberspace computing, banking, fund transfer systems, aerospace, avionics, and any distributed AI application, call for obvious needs to have techniques of designing, implementing, and proving that AI systems are built with proper means of providing privacy. There has to be techniques to design autonomous agents with safe parameters and comfort zones, while rendering transgressor processes ineffective. Autonomous safe multiagent computing can be a goal defined by formal means in our papers. The problems of *privacy and security* for computer systems are not new. It has been an occupation for many computer scientists and engineers, ever since the design of the second generation of operating systems, to demonstrate that a system does provide the two pillars. There were, and are today, numerous projects to define what the two pillars are, how to design systems that could have proper methods for them, and how to prove that systems had the functionality needed to provide integrity. Thus at the present time AI computing and the implemented systems lack the two pillars of their own rights and have to rely on what an operating system does. That is, of course, inadequate for multiagent and distributed AI systems. The existing methods are not applicable to the systems that are to be designed with some new and necessary trends for AI.

4. CYBERDOCKING VIA MOBILE INTERFACES

Cyberdocking modules (Nichols and Lekkas 2002) are interfaces to Web browsers which can manage cyberspace interactions to be only known and defined functional activities to the Windows host computer and user environment. A cyberspace docking facility design and network Interfaces capable to select only intended functionality from Cyberspace without side-effects to the Windows user's computing environment integrity. Proven techniques for design and implementation of AI software agents are to be applied to design windows interfaces with conventional windows technology. The basic design concepts were defined by the

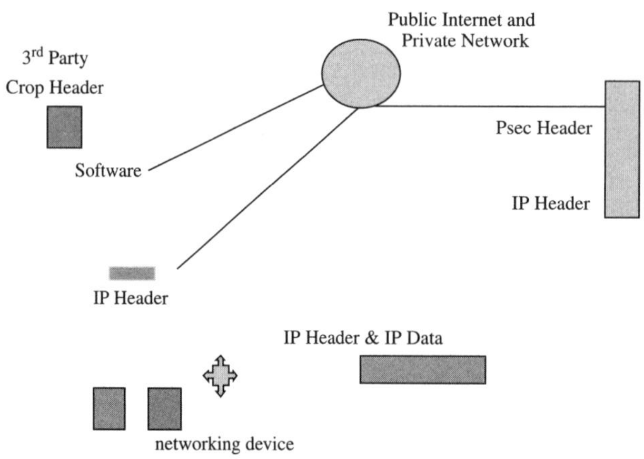

Figure 1. Typical psec usage.

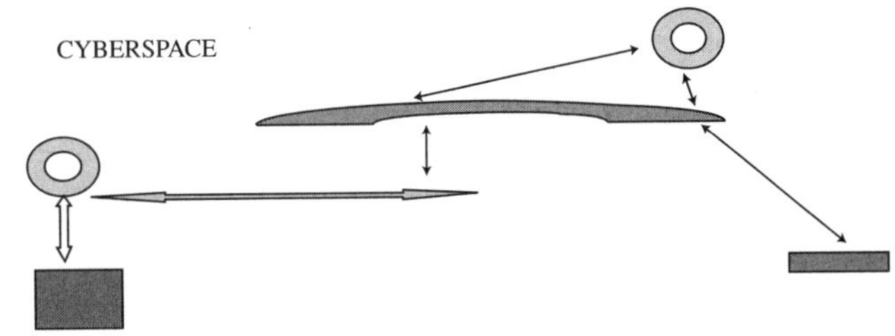

Figure 2. Virtual VPN.

authors since (1993–1997 publications), embedding basic privacy and integrity for windows applications. CyberIntelligent, i.e., being Cyber-aware to Cyberspace functionality and its actions onto to the Cyberdocking and the user environment is essential. Docking facilities apply interceptors (Nichols and Lekkas 2002), which are multiagent modules running with the Cyberdocking module intended to intercept unexpected functions from the Cyberspace or from an Internet operation, e.g., ftps. The principles have been defined with objects and multiagent AI for designing systems with software agent principles with agent functions, objects, and OS kernels. Obligations, sharing resource and necessities implied functionality could be designed. The designs have been proven to be superimposable onto the existing OS's. There are specific designs on the auhtor's outstanding patent disclosures to design docking as virtual WWW launches via distant instantiations where Aglet computations can be invoked limited to virtual Cyberballs.

4.1. Mobile Agents and WWW

A Java aglet is similar to an applet in that it runs as a thread (or multiple threads) inside the context of a host Java application. To run applets, a Web browser fires off a Java application to host any applets it may encounter as the user browses from page to page. That application installs a security manager to enforce restrictions on the activities of any untrusted applets. To download an applet's class files, the application creates class loaders that know how to request class files from an HTTP server. An aglet requires a host Java application to be running on a computer before it can visit that computer. When aglets travel across a network, they migrate from one aglet host to another. Hosts upload aglets through class loaders that know how to retrieve the class files and state of an aglet from a remote aglet host. The techniques are applied to mobile computing (Nourani and Moudi 2001) with aglets to obtain specific sound mobile computing privacy and security criteria. The basic techniques are foundational to firewall designs. The intelligent syntax and object computing paradigm is applied to Aglet computing as an agent Java variant with all the Web logical basis developed in our papers since the last few years.

4.2. Authentication Header

Authentication Header (AH) allows IPSEC peer systems to ensure the integrity of exchanged messages. The typical interfaces and psec headers are depicted. Storing and matching WWW

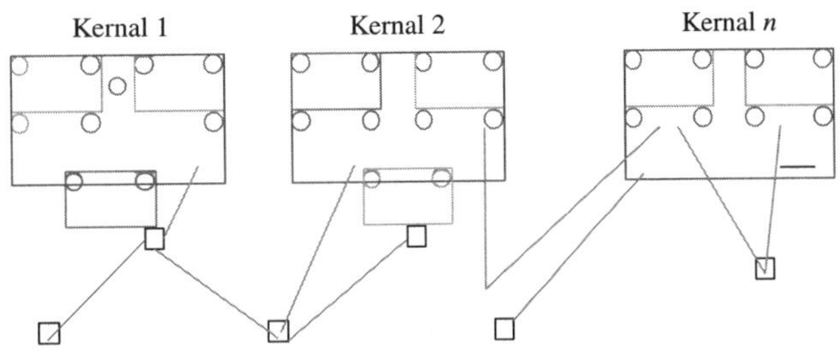

Figure 3. Multiagent multi-kernel design.

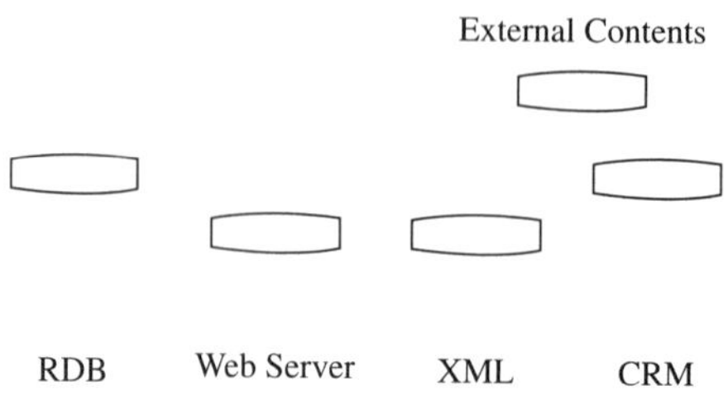

Figure 4. The content store.

contents via knowledge-bases and databases might be viewed as satisfying a goal by getting at relevant data which instantiates a goal. The goal formula states what relevant data is sought. The techniques can be applied to implement discovery planning. In planning with generic model diagrams, G-diagrams (Nourani 1991) is as follows: that part of the plan that involves free Skolemized trees is carried along with the proof tree for a plan goal. The idea is that if the free proof tree is constructed then the plan has a standard model in which the goals are satisfied. The model is the initial model of the AI world for which the virtual trees are presented.

4.3. Virtual Generic Virtual Private Network and WAP

A virtual WAP–VPN is a module designed with mobile agents and objects, for example Aglets, to cast a virtual environment on the cyberspace or a predesignated site in order to grant access and engagements to specified host private and secure functions. Virtual kernels can be invoked via WAP–VPN. The specific designs on Cyberballs and docking modules are subjects on the author's patent disclosures. Cyberdocking modules are might invoke Cyberballs in there design. There is an important distinction in that docking is to manage a mutual engagement with specific goals between two sites at a designated virtual address at cyberspace to cooperate on a specific task. Cyberballs allow computations with the provided functions. WAP–VPN insolate the virtual access to a host site's functions from intrusions from unauthenticated agents. Figure 2 depicts the basic design where the double circulars are the virtual mobile private and secure planet interfaces from Nourani and Moudi (2001). The square is the local interface base, for example a PC or cellphone. The WiFi concept 'Bubble' stated in the Wired magazine 2003 since are not distant from what can be realized here. The kernel segment is selected and cast from a multiagent multikernel design from Nourani and Moudi (2001), Nourani (1998a) as depicted.

5. KR, KB, AND VISUAL WWW INTERFACES

Model diagrams allow us to model-theoretically characterize incomplete KR (Nouani and Lieberherr 1986). To key into the incomplete knowledge base we apply generalized predictive

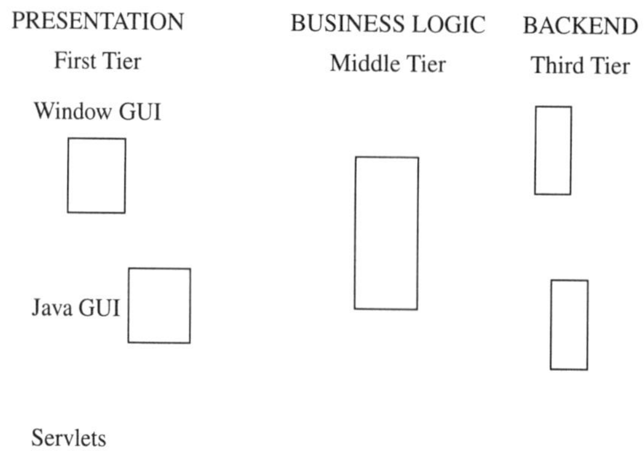

Figure 5. J2ME and content interfaces.

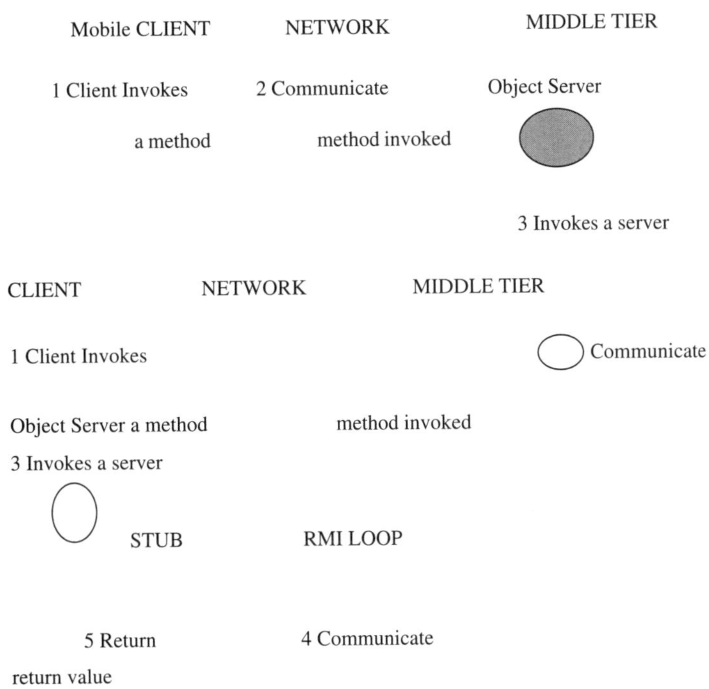

Figure 6. The RMI loop.

diagrams whereby specified diagram functions a search engine can select onto localized data fields. The predictive model diagrams (Nourani 1995) could be minimally represented by the set of functions $\{f_1, \ldots, f_n\}$ that inductively define the model. Data discovery from KR on diagrams might be viewed as satisfying a goal by getting at relevant data which instantiates a goal. The goal formula states what relevant data is sought. We propose methods that can be applied to planning (Nourani 1991) with diagrams to implement discovery planning. In planning with G-diagrams that part of the plan that involves free Skolemized trees is carried along with the proof tree for a plan goal. Computing with diagram functions allows us to key to active visual databases with agents. The incomplete knowledge modelling is treated with KR on predictive model diagrams.

Model discovery at KB's are with specific techniques defined for trees. Model diagrams allow us to model-theoretically characterize incomplete KR. To key into the incomplete knowledge base we apply generalized predictive diagrams whereby specified diagram functions a search engine can select onto localized data fields. The predictive model diagrams (Nourani 1995, 1999a) could be minimally represented by the set of functions $\{f_1, \ldots, f_n\}$ that inductively define the model. Data discovery from KR on diagrams might be viewed as satisfying a goal by getting at relevant data which instantiates a goal. The goal formula states what relevant data is sought. The first author has presented techniques that can be applied to planning with model diagrams to implement discovery planning. In planning with G-diagrams that part of the plan that involves free Skolemized trees is carried along with the proof tree for a plan goal. Computing with diagram functions allows us to key to active visual databases with agents.

5.1. Content Management

To design interfaces with systems we have to provide content management interfaces as presented in Chapter 11 can be applied with the J2ME server and the WAP–VPN. The stub implements an interface with the same business methods as the object itself, but the stubs methods do not contain business logic. The data tier consists of one or more databases and may contain data-related logic in the form of stored procedures.

EXERCISES

1. Where do Crypto Assets reside in WAP designs?
2. What distinguishes the virtual VPN design here from a static VPN?
3. How do you apply the WAP–VPN design here to realise WiFi 'bubbles.'
4. *Cybersafety and Aglet Life*: A necessary condition to attain cybersafety with mobile agent computing with aglets is that the Aglet life style be sound with respect to certain basic design goals. Can you state your design goals and what simplifying hypothesis you are applying.
5. State examples where Psec's are applied on interfacing with Aglet's and authenticated on Cyberballs.

REFERENCES

1. F. M. T. Brazier, B. Dunin-Keplicz, N. R. Jennings, and J. Treur, "Formal Specification of Multi-Agent Systems: A Real-World Case" (V. Lesser, Ed.), pp. 25–32. Proc. of the First International Conference on Multi-Agent Systems. ICMAS-95, MIT Press, Cambridge, MA, 1995.
2. M. R. Genesereth and N. J. Nilsson, "Logical Foundations of Artificial Intelligence." Morgan-Kaufmann, 1987.
3. J. Hagel and A. G. Armstrong, "Net Gain—Expanding Market Through Virtual Communities." Harvard Business School Press, Boston, Massachussetts, 1997.
4. R. K. Nichols and P. C. Lekkas, "Wireless Security." McGraw Hill, 2002.
5. N. J. Nilsson, "Searching, Problem Solving, and Game-Playing Trees for Minimal Cost Solutions" (A. J. Morell, Ed.), Vol. II, pp. 1556–1562. IFIP, North-Holland, Amsterdam, 1969.
6. C. F. Nourani, "Planning and Plausible Reasoning in AI," pp. 150–157. Proc. Scandinavian Conference in AI, IOS Press, Denmark, 1991.
7. C. F. Nourani, "Fields of Gold." 1993a.
8. C. F. Nourani, "Abstract Implementations Techniques for AI Systems by Computing Agents—A Conceptual Overview." Proc. SERF-93, Orlando, FL, 1993b.
9. C. F. Nourani, "Slalom Tree Computing—A Computing Theory for Artificial Intelligence," Vol. 4, No. 4. AI Comm., 1994.
10. C. F. Nourani, "Free Proof Trees and Model-Theoretic Planning and Nonderministic Algebraic Plan Computation with G-Diagrams." AISB, Sheffield, England, 1995.
11. C. F. Nourani, "Cyberspace Docking Modules and Interceptors." Based on a venture proposal written to Microsoft, 1997.
12. C. F. Nourani, "Designing Fault Tolerant Systems with Multiagent AI." FTCS98, Munich, Germany, 1998a.
13. C. F. Nourani, "Intelligent Languages a Preliminarily Syntactic Theory." Proc. 23rd International Mathematical Foundations Computer Science, Silesian University, Institute of Computer Science, Grammatics, Springer, 1998b.
14. C. F. Nourani, "Agent Computing, KB For Intelligent Forecasting, and Model Discovery for Knowledge Management." AAAI—Agent Based Systems in the Business Context, Orlando, Florida, 1998c.
15. C. F. Nourani, "A Parallel Functional Modular Language With Agents and Distributed Objects," Brief, IFL Wkshp Parallel Functional Languages, London, 1998d.
16. C. F. Nourani, "KR and Model Discovery with Predictive Logic." AI Conference, AKA Wkshop, Sydney, 1999a.
17. C. F. Nourani, "Multiagent AI Design a New Emerging Software Engineering Trend," Vol. 12, No. 1, pp. 37–42. Engineering Applications AI—EAAI, Elsevier, 1999b.
18. C. F. Nourani, "Intelligent Multimedia, New Computing Techniques, Design Paradigms, and Applications" Treeless Press, Berkeley, CA, 2000.
19. C. F. Nourani and K. J. Lieberherr, "Data Types, Direct Implementations, and Knowledge Representation," Vol. II, pp. 233–243. Proc. 19th HICSS. Honolulu, Hawaii, 1986.
20. C. F. Nourani and R. M. Moudi, "CyberSignatures, the Cyberspace Interfaces and Cooperative Computing." ISKDE, Computing Science, The University of Liverpool, UK, 2000.
21. C. F. Nourani and R. M. Moudi, "Intelligent Interfaces, Agent Process Models, and Mobile Computing." ICAI, CSREA Press, Las Vegas, 2001.
22. C. F. Nourani and R. M. Moudi, "Intelligent Cyberspace Interfaces and WAP Generic Computing." IKS, St. Thomas, Virgin Islands, 2002.
23. F. J. Riggins and H. S. Sue Rhree, *Communications Association Computing Machinery* 41, 88 (1998).
24. The Mobile Object Papers, Vol. 41, No. 10. ACM 1998.
25. The IBM's aglets home page http://www.trl.ibm.co.jp/aglets
26. Gio Wiederhold, "The Roles of Artificial Intelligence in Information Systems," pp. 38–51. In Ras Methodologies for Intelligent Systems, Lecture Notes in AI, Springer, 1991.

Index

ISBN: 1-58883-037-3/$35.00

Intelligent Multimedia Computing Science
By Cyrus F. Nourani
Pages: 187–190

E

F

G

H

I

J

K

T

V

W